The Beginning Of Days

The Beginning Of Days

By

Reggie Frasier

Copyright *CE 2016* ©

All rights reserved.

No part of this book may be reproduced, in any form or by any means, without permission in writing from the publisher.

**All biblical quotations taken from
The Authorized King James Version of the Bible, 1611**

Books by Reggie Frasier:

The End of Days

The Great Tribulation and the War in Heaven

The Apotheosis of the Apocalypse

Daniel the Prolific Prophet

The Beginning of Days

Published in the U.S.A.

by

𝓣 𝓗

Thaddeus House
PO Box 76
Grayson, GA 30017

The Beginning of Days

Contents

Introduction	11
The Creation	13
The Creation Gap	16
The Career of Satan	26
The Reconstruction	47
The Sons of God	82
The Giants in the Earth	104
Scientific Evidence	121
The Destruction of the Earth	123
Creation	135
Three Theories of the Beginning	140
The Theory of the Eternal Universe	140
The Big Bang Theory	141
The Theory of Energy Conversion to Mass	142
The Agent of Mass Destruction and Its Effects on the Solar System	146
The Minor Bodies	147
The Major Bodies	160
The Sun	160
Mercury	164
Venus	165
Earth	167
Mars	174
Jupiter	178
Saturn	185
Uranus	196
Neptune	202
Pluto/Charon	207
The Ages of the Earth	211

Earth's Major Extinctions 215
Geologic Evidence on the Continents 232
    Asia 232
    Europe 235
    South America 237
    North America 241
    Africa 247
    Antarctica 254
Biblical Proofs 256
Continual Catastrophes Down Through the Ages 258
    From Adam to Moses 258
    From Moses to the Wandering 274
    From The Wandering to the Conquest 292
    From the Conquest of Canaan to the Judges 298
    From Judges to Solomon 302
    From Solomon to Elisha 307
    From Elisha to Jesus 313
    From Jesus to Today 321
Conclusion 334
About the Author 335

The Beginning of Days

Genesis 1:1; "In the beginning, God created the heaven and the earth."

John 1:1-3; "In the beginning was the Word, and the Word was with God, and the Word was God. ² The same was in the beginning with God. ³ All things were made by him; and without him was not any thing made that was made."

Dedication:

To my wife, Carolyn,
who inspires me to
greater and better
endeavors every day.

Introduction

The purpose of this work is to identify the meaning inherent in the many obscure and misunderstood scriptures in ***The Book of Genesis;*** and other select sections of the Old Testament. The earlier stories of mankind on the Earth are sometimes hemmed in by a nearly impenetrable and misty fog obscuring the reliability of their exact meaning and therefore the true nature of the events being described. It takes an open mind and a discerning spirit to accurately understand what the more ambiguous depictions of events are truly describing.

A truly thorough reading of these scriptures reveal what was happening on the Earth of a celestial nature, and exposes truly monumental catastrophism affecting the Earth and all of its inhabitants during this devastating time in man's history.

It reveals a planet sized body careening down through the Solar System, affecting each and every member of the ancient planetary realm, and conversely, the beings on the third planet from the Sun, the Earth.

This Agent of Mass Destruction caused much damage and left an indelible impression on man as he began to emerge from his darkest age, the stone-age, and began recording his history, recording events as he perceived them, in his own superstitious and religiously inclined mind.

The early books of the Bible reflect this state of affairs, man recording what he heard, saw and experienced from his own frame of reference, not necessarily a pure scientific endeavor. This has been passed down to prospective generations through the Bible and each culture's mythologies and histories. The purpose of this study will be to reconcile science with scripture by interposing this scenario, the Agent of Mass Destruction careening through the Solar System, It also allows a good chance to study God's Word, thereby developing a better relationship with Him and to understand what truly was occurring in those vague first historical epochs.

This work presents a captivatingly different and alluringly refreshing interpretation of what those observations meant and exposes the Agent of Mass Destruction for what it did to the Solar System and though this is not the acceptingly conventional interpretation of what the patriarchs were telling us, it does answer the most questions concerning what those early writings actually signify.

The author covered the events of the "End of Days" in his previous books and now he uses the same eloquent perspicuity to scrutinize the earlier elusive writings composing the Bible.

It is the author's hope you will benefit from all his writings.

The Creation

In The Beginning

***Genesis 1:1;** "In the beginning God created the heaven and the earth."*
***John 1:1-3;** "In the beginning was the Word, and the Word was with God, and the Word was God. The same was in the beginning with God. All things were made by him; and without him was not any thing made that was made."*

Crisis of Faith

Every time archaeologists in the field find fossilized dinosaur bones which date around 65 million years ago or older, or when astronomers find stars or galaxies with ages of 15 to 20 billion years, *(Scientific Creationism)* and this is contrasted with what some men claim was the creation as accomplished by God and described in the first book of the Bible, **The Book of Genesis;** *(Biblical Creationism)* a normal person will have a crisis of faith.

He will either have a crisis of faith in science, which he believes is basing its conclusions on inaccurate data, or he will have a crisis of faith in God, the Author of the Bible, because he will believe the Bible is inaccurate in what it describes.

He has this crisis because science contends the creation occurred over 13.8 billion years ago and the Bible has been interpreted to suggest the creation occurred a mere 6,000 years ago.

This 6,000 year figure was first proposed by James Ussher, *CE 1581–1656* Archbishop of Armagh, Ireland in *CE 1625*, when he established his meticulous analysis of the genealogical time spans of the ancient patriarchs of the Bible, and arrived at the date *4,004 BCE* as the date of the biblical creation. Archbishop Ussher became very precise in his calculations. He had the temerity to pinpoint the moment as *"6 o'clock p.m. on the evening of 22 October, 4004 B.C."* in the Julian calendar even though the Gregorian calendar had already been introduced *CE 1582*.

The certainty of his own calculations and the pressing need of the day to establish proof of a young Earth in opposition to the billions of years of age scientists were proposing at the time, caused the Prelate of All Ireland's attempt at dating to go unquestioned in theological circles.

Men would have done well to heed God's warning about setting dates, **Matthew 24:36;** even though God was warning against the timing of prophecy, especially the Second Coming. **The Book of Genesis;** is just as uncertain as prophecy in many instances.

Archbishop Ussher proposed the date was for the creation, when it will be shown to be the date of the expulsion of Adam from the Garden of Eden.

This disparity is what has fomented the crisis of faith in many believing people. This and the misunderstood creation of man, as will be explored in this work.

Is there a way to alleviate this crisis of faith? Is there a way to reconcile science and the Bible?

This thesis proposes a thorough study of the whole counsel of God's Word can resolve these discrepancies.

Study of the Scriptures

II Timothy 2:15; *"Study to show thyself approved unto God, a workman that needeth not to be ashamed, rightly dividing the word of truth."*

Believers are instructed to study God's Word in our endeavor to know Him in His fullness.

There are people who boast they have read the scriptures all the way through many times in their life, and this is good in the sense they are at least reading God's Word, and they have the blessing inherent in the reading the Word of God. ***Revelation 1:3;***

However, God wants us to study *(to take apart and dissect for meaning)* His Word. For it is through studying, the true meaning of the verses, and therefore, the true nature of God, emerges, for the entirety of God and the fullness of His being is enclosed and revealed in the scriptures.

Matthew 13:52; *"Then he said unto them, Therefore every scribe which is instructed unto the kingdom of heaven is like unto a man that is a householder, which bringeth forth out of his treasure things new and old."*

An effective student in the study of His Word, will be like this householder, bringing new things out of the treasure, as well as old.

When studying the Word of God, with each area of study, one should learn something new, as well as being reminded of the old. A truly effective and obedient student will always be progressing toward a relationship with the Author.

This, after all, should be ones only goal.

This study will begin at the beginning and start by examining what the Word of God actually says about the creation.

The Creation Gap

Scientists have estimated the creation of the Universe was accomplished by a Big Bang approximately 13.8 billion years ago. The Bible declares the creation happened at the beginning, indicating time also started simultaneously with creation. Based on what was discussed above about the timing of the creation in biblical circles, *4004 BCE,* there seems to be quite a gap between these two incongruent versions.

The Creation Gap, also known as The Gap Theory, claims a large gap of time existed between the first two verses of the Bible. The scriptures used to expound The Creation Gap, are as follows: **Genesis 1:1-2; Isaiah 45:18; Jeremiah 4:23-28; II Peter 3:5-7;**

To understand the Creation Gap correctly, and to analyze these scriptures according to what is actually being said, it is necessary to get the Hebrew renderings of the words used in these scriptures.

Genesis 1:1-2; "In the beginning God created *Hebrew: bara:* בָּרָא*, to make something new, something not there before, from materials not present before,* ***the heaven and the earth. And the earth was*** *Hebrew: hayah:* הָיָה *to come to pass, to become* ***without form****, Hebrew: tohuw:* הוּא *a desolation* ***and' void;*** *Hebrew: bohuw:* הוּ *to be empty* ***and darkness was upon the face of the deep.*** *Hebrew: tehom:* תְּהוֹם *the abyss, a great expanse of water, a subterranean sea* ***And the Spirit of God moved upon the face of the waters."***

The combination of these two verses at the start of the scriptures has led to a grave and momentous misinterpretation of the Word of God, a critical misunderstanding promulgated throughout the Christian era.

It seems on the surface to be saying, when God created the Heaven and the Earth, the Earth was a chaotic emptiness. If one takes into account the whole counsel of the scriptures, and realizes the scriptures are the inspired Word of God, **II Timothy 3:16;** one will see the Word of God does not say this at all.

First, it goes against everything known and understood concerning God and His perpetual perfection. When God creates something, it is perfect. **Deuteronomy 32:4;** If one takes these verses on the surface, they seem to be saying God created the Heaven and the Earth in a state of total chaos.

Second, there is a discrepancy about the Hebrew rendering of the word 'was', *Hebrew: hayah:* הָיָה which means 'became' many more times in translation than it means 'was'. The word *Hebrew: hayah:* הָיָה is translated 'was' many times in scripture, and those renderings seem justified, but there is no indication in the syntax of this verse, it should mean 'was' here, rather, 'became' seems to be the desirable translation here to hold coherence in the Hebrew. And also to hold coherence in what is known of God, simply because He would not create something in a state of chaos, instead He would create something perfect. Since it is in the state of chaos in **Genesis 1:2;** it would seem to indicate something happened to make it *"without form and void"*.

There is another verse clearly stating God did not create the Earth without form.

Isaiah 45:18; "For thus saith the LORD that created the heavens; God himself that formed the earth and made it; he hath established it, he created it not in vain, *Hebrew: tohuw:* הוּא *a desolation,* ***he formed it to be inhabited: I am the LORD; and there is none else."***

Just by looking at this one other alternate scripture, it is clear, God did not create the Earth in a chaotic condition. *Hebrew: tohuw:* תֹהוּ He "...*formed it to be inhabited...*". Yet the Earth is in the state of disorder and confusion *Hebrew: tohuw:* תֹהוּ in the **second verse of Genesis One;**.

Hebrew: hayah: הָיָה

Hebrew: hayah: הָיָה *to exist ,to become, come to pass, was*

According to Strong's Exhaustive Concordance of the Bible, in scripture the word *Hebrew: hayah:* הָיָה is translated in alphabetical order as; accomplished, be, became, become, been, cause, caused, come to pass, cometh, continue, continued, fainted, follow, followed, happened, had, hast, hath, have, lasted, pertained, pertaineth, quit, required, was,

Looking at the Hebrew word *Hebrew: hayah:* הָיָה which is translated as 'was' in the **second verse of Genesis One;**, and realizing, more times than not, the word is translated 'became' in scripture, it follows, a possible rendering of the second verse of the Bible could be, "***And the earth became without form and void...***" which would give it an entirely different meaning from the one now commonly held, and the one used by the translators of the King James Authorized Version of the Holy Scriptures, and would justifiably, decisively and reasonably reinforce the insightful argument for the Creation Gap.

If God did not create the Earth as a chaotic emptiness *Hebrew: tohuw:* תֹהוּ in what manifestation was it created?

It states in ***Isaiah 45:18;*** "...*he formed it to be inhabited.*" A planet in a state of total chaos cannot be inhabited, and this verse says when God created it, it was ready for habitation. It wasn't created in a state of total chaos and pandemonium, rather it was made perfect and flawless, ready to be inhabited. And inhabited it was, with dinosaurs, various flora, fauna and marine life, and even men, long before Adam was finally created in **Genesis Two;**, and scientists are finding the fossilized remains of this abundant life all over the Earth today.

Jeremiah's Vision

The men mentioned above were the inhabitants of the Old Earth, before the First Great Flood. This First Great Flood was not the flood of Noah's day, but one of far greater magnitude, at an earlier time, one accounting for the waters on the face of the deep, **Genesis 1:2;** one which resulted in God having to remake everything on the Earth, an act this treatise refers to as the Reconstruction, a flood glimpsed far better in another scripture.

Jeremiah 4:23-28; "*I beheld the earth, and, lo, it was without form*, Hebrew: tohuw: תֹהוּ *and void;* Hebrew: bohuw: בֹהוּ *and the heavens, and they had no light.* (Darkness was upon the face of the deep. **Genesis 1:2;**) *I beheld the mountains, and, lo, they trembled, and all the hills moved lightly. I beheld, and, lo, there was no man,* Hebrew: adam: אָדָם *ruddy and all the birds of the heaven were fled.* Hebrew. nadad: נָדַד *to drive away* (it is just becoming clear the nearest living relatives of the dinosaurs alive today are the birds) *I beheld, and, lo, the fruitful place was a wilderness, and all the cities thereof were broken down at the presence of the LORD, and by his fierce anger.* (Anger kindled in the LORD by the fall of his anointed cherub Lucifer/Satan. **Ezekiel 28-12-19;**) *For thus hath the LORD said, The whole land shall be* *desolate; yet will I not make a full end. For this shall the earth mourn, and the heavens above be black: because I have spoken it, I have purposed it, and will not repent, neither will I turn back from it."*

The Bible has much to say about the state of the Earth as the prophet Jeremiah viewed it in his lucidly confirming vision.

This vision had to be of the Earth as it stood in **Genesis 1:2;**, because then was the only time in history or prehistory the Earth was 'without form and void' *Hebrew: tohuw:* תֹהוּ and *Hebrew: bohuw:* בֹהוּ just as Jeremiah stated it was as he was having his vision. He also said there were fruitful places now turned into wilderness, and the cities were broken down or destroyed. This would indicate habitation in some form by man, as God confirms in **Isaiah 45:18;** *"...he formed it to be inhabited."*

The Old World (Earth)

II Peter 3:5-7; "For this they willingly are ignorant of, that by the word of God the heavens were of old, *(*in the beginning) ***and the earth standing out of the water*** (inhabited) ***and in the water:*** (covered by the First Great Flood) ***Whereby the world that then was,*** (the Old Heaven and Old Earth which were created in the beginning) ***being overflowed with water, perished:*** (at the fierce anger of the LORD) ***But the heavens and the earth, which are now,*** (meaning the current Heaven and Earth are different in some way from the ones of old, or as they were when they were originally created) ***by the same word are kept in store, reserved unto fire against the day of judgement and perdition*** (the Great White Throne Judgment) ***of ungodly men."***

This clearly declares the Heavens, at the beginning (of old), were created by the Word of God, (Jesus, ***John 1:1-3;***) and the Earth, in the beginning (of old), was standing *"...out of the water..."*, then *"...in the water..."*, at the fall of Satan ***Isaiah 14:12-20; Ezekiel 28:13-19;***, the First Great Flood, and it affirms the world *"...that then was..."* perished by the overflowing of the vast quantities of water.

It declares the Heaven and the Earth which are now, distinguishing this present Heaven and Earth from the one of Old, are reserved for fire ***II Peter 3:10-12;*** and the day of judgement.

This differentiation of the Old Heaven and Old Earth from this present Heaven and Earth is further alluring proof of the Creation Gap. ***Genesis 1:1 and 1:2;***

Noah's Flood

Many biblical scholars consider the verses in ***II Peter 3:5-7;*** as referring to the flood of Noah's day, but this thesis contends this is a description of a greater flood before the days of the Reconstruction, ***Genesis 1:2;***, the First Great Flood, because these verses refer to a New Heaven and a New Earth after the ubiquitous, devastating overflow of water completely destroyed the Old Earth.

In the flood of Noah, the whole Earth was overflowed with unescapable water and every living creature was destroyed, those who lived on the Earth's land surface, except those who were preserved in the Ark. This widespread destruction did not encompass the marine life, and the plant life was not utterly destroyed, however, in the First Great Flood, even the plant life was entirely obliterated and necessitated a requisite reforming in the Reconstruction. ***Genesis 1:11-12;***

Also, in the flood of Noah's days, the Heavens did not need a renewing, as it did after the First Great Flood, when the LORD destroyed the Earth, and made the Heavens black, *(darkness)* **Jeremiah 4:28;** at His presence and by His fierce anger. **Jeremiah 4:26;** In Noah's day, after the flood of those days, the Heavens above the watery Earth had exactly the same appearance as before the flood, the constellations may have looked slightly different, since the Earth was moved slightly on its axis, now inclined at twenty three and a half degrees (23.5°) to the plane of the ecliptic, but the Heavens were still there and the planets were still in the same order, meaning the Earth did not change its place in the Solar System, after Noah's flood. However, the First Great Flood of **Genesis 1:2;** was accompanied by a changing of the place of the Earth in relation to the Sun, the stars and the planets. This indicates the Earth was probably moved out of its regular rotation, moving to a new rotation, the one seen today, and maybe even exchanging its North/South geographical and magnetic poles during this supremely monumental event. It is also probably true the Earth changed its place in the Solar System sometime during this First Great Flood. It was probably much further out near the planet Jupiter before the First Great Flood and the devastation wrought by the LORD and His fierce anger at this far distant time. This is what is described by the apostle Peter. **II Peter 3:5-7;** This is also what the prophet Jeremiah saw in his perplexing vision. **Jeremiah 4:23-28;**

This is also further confirmation of the Creation Gap.

Finding Room for the Dinosaurs

Within this Creation Gap is located all the empirical evidence which scientists, archaeologists, anthropologists, botanists and geologists are now beginning to discover about this 4.5 billion year old planet Earth embodied in all the dinosaur bones and other fossil evidence being unearthed from the planet.

The biggest single argument against the veracity of the scriptures being presented in the scholarly world today is the argument the scriptures leave no room for the dinosaurs and all the ages of the Earth. Now, this argument falls flat and utterly fails. There is more than enough time and space for all the bones scientists are digging out of the Earth, and still allow the Word of God to maintain its integrity.

This time and space and this integrity in the Word of God are explained by the Creation Gap.

Replenish the Earth

This thesis will now introduce another substantial support in the scriptural narrative to reinforce the Creation Gap.

When God created the man on day six of the Reconstruction, he told them to **"Be fruitful and multiply, and replenish the earth." Genesis 1:28;**

The very meaning of the word 'replenish' denotes an act which is re-doing something previously done. It is from the root words *Re - to do again and plenus - fill.*

It is hard to replenish an Earth which was not filled in the first place. This command to man is another verification for the fact there were inhabitants of the Old Earth, before the First Great Flood and the total destruction of the Old Earth in long ago ages. It is also a validation of the Creation Gap. It seems clear God is telling man to fill the Earth again, implying at some time before man was created, (day six) it was full, populated. This could only be referring to a population which filled the Earth before man was created (day six).

It is the same command He gives to Noah and his family upon their debarkation from the Ark. ***Genesis 9:1;*** God chose this word to relay His meaning to refill the Earth, as it once was full. Instead of saying ***"Be fruitful, multiply and fill the earth,"*** God says, ***"Be fruitful, multiply and replenish the earth" Genesis 9:1;*** It is stated with assurance the world before the flood of Noah was indeed full and populated, in the same way it is stated the world existing before the creation of man (day six), and before the Reconstruction of ***Genesis One;*** was full and populated.

This is more evidence God gives to the actual situation of the Earth in ***Genesis One;***, and He did it this way to encourage study in His Word for the hidden things, things only revealed through careful scrutiny. ***II Timothy 2:15;***

Now, it is time to examine the reason the original Earth was destroyed by the LORD, and by His fierce anger.

The Career of Satan

The Facts:

The scriptures reveal an abundant amount of information about our arch enemy, and though surely it is not everything known about this immoral and dissolute being, it should be more than ample data to judge the "*...wiles of the devil...*" **Ephesians 6:11;** enough for any devout believer to adequately undertake his lifelong struggle against him.

As unpleasant a subject as this is, this thesis will quickly dispense with it here. If the discerning reader would like a supplementary study of this consummate evil in our midst, he is encouraged to read and study the author's other book, **"The Great Tribulation and the War in Heaven"**, where there is a major section on this wicked and despicable fallen angel and the deleterious effect his fall had on the entirety of creation, especially man and the other hapless denizens of the Earth.

There are two major important blocks of scripture identifying and exposing in detail this reprehensible and nefarious being. They are *Isaiah 14:12-20; and Ezekiel 28:11-19;.*

The Old Testament

Isaiah 14:12-20; "How art thou fallen from heaven, O Lucifer, son of the morning! how art thou cut down to the ground, which didst weaken the nations! For thou hast said in thine heart, I will ascend into heaven, I will exalt my throne above the stars of God: I will sit also upon the mount of the congregation, in the sides of the north: I will ascend above the heights of the clouds; I will be like the most High.⁵Yet thou shalt be brought down to hell, to the sides of the pit. They that see thee shall narrowly look upon thee, and consider thee, saying, Is this the man that made the earth to tremble, that did shake kingdoms; That made the world as a wilderness, and destroyed the cities thereof; that opened not the house of his prisoners? All the kings of the nations, even all of them, lie in glory, every one in his own house. But thou art cast out of thy grave like an abominable branch, and as the raiment of those that are slain, thrust through with a sword, that go down to the stones of the pit; as a carcase trodden under feet. Thou shalt not be joined with them in burial, because thou hast destroyed thy land, and slain thy people: the seed of evildoers shall never be renowned."

and

Ezekiel 28:11-19; "Moreover the word of the LORD came unto me, saying, Son of man, take up a lamentation upon the king of Tyrus, and say unto him, Thus saith the LORD God; Thou sealest up the sum, full of wisdom, and perfect in beauty. Thou hast been in Eden the garden of God; every precious stone was thy covering, the sardius, topaz, and the diamond, the beryl, the onyx, and the jasper, the sapphire, the emerald, and the carbuncle, and gold: the workmanship of thy tabrets and of thy pipes was prepared in thee in the day that thou wast created. Thou art the anointed cherub that covereth; and I have set thee so: thou wast upon the holy mountain of God; thou hast walked up and down in the midst of the stones of fire. Thou wast perfect in thy ways from the day that thou wast created, till iniquity was found in thee. By the multitude of thy merchandise they have filled the midst of thee with violence, and thou hast sinned: therefore I will cast thee as profane out of the mountain of God: and I will destroy thee, O covering cherub, from the midst of the stones of fire. Thine heart was lifted up because of thy beauty, thou hast corrupted thy wisdom by reason of thy brightness: I will cast thee to the ground, I will lay thee before kings, that they may behold thee. Thou hast defiled thy sanctuaries by the multitude of thine iniquities, by the iniquity of thy traffick; therefore will I bring forth a fire from the midst of thee, it shall devour thee, and I will bring thee to ashes upon the earth in the sight of all them that behold thee. All they that know thee among the people shall be astonished at thee: thou shalt be a terror, and never shalt thou be any more."

Here are the facts these scriptures reveal:

He is a created being of the class termed angels. ***Ezekiel 28:13 &15;***

He was perfect in beauty and full of wisdom. ***Ezekiel 28:12 & 17;***

Before his fall, he was God's ***"...anointed cherub that covereth...",*** a reference to him as the one in charge of all of God's creation. ***Ezekiel 28:14***; God's trusted manager of all the wondrous works of His hands.

His position of authority, before his untimely fall, was because God appointed him to this prominent and important position. ***Ezekiel 28:14;***

Before his fall, he had as his covering, his raiment, every precious stone and gem known to be of great value. ***Ezekiel 28:13;***

At his creation, he was prepared with the workmanship of tabrets and pipes, or in other words, he was a being physically and spiritually invested with musical instruments and ability as part of his entire makeup. Music was integral to his being. *Ezekiel 28:13*;

He walked upon the holy mountain of God. *Ezekiel 28:14*;

He walked up and down in the midst of the stones of fire. *Ezekiel 28:14*;

He was absolutely perfect in his ways from the first day of his creation, and he was merely a created being, until iniquity was found in him. *Ezekiel 28:15;*

When he sinned, his iniquity came like a blazing inferno from within himself and grew to such a great conflagration it consumed him completely being the only trait remaining by which he would forever thereafter be defined. *Ezekiel 28:18;*

He fell because he was filled with overwhelming determination to be like the Most High God, *Isaiah 14:14;*

He fell because he was filled with odious violence and sinned, *Ezekiel 28:16;* God equates violence with sin.

He fell because he was arrogantly lifted up and infused with vainglorious vanity by reason of his amazing beauty, *Ezekiel 28:17;*

He corrupted his sagacious wisdom because of his blinding and irresistible brightness. *Ezekiel 28:17*;

He maintained dominion over the Earth, and all its doleful inhabitants before his fall, *Isaiah 14:30;* This is further proof the Old Earth was prolifically inhabited.

God lays the blame of the absolute destruction of his land (the Old Earth) and the extensive slaying of his people (the previous inhabitants of the Old Earth) at Satan's door. *Isaiah 14:20*;

He existed before the beginning, because at the beginning, when God was laying the foundations of the Earth, Satan was present singing as one of the bright and morning stars, in unity with all of the other angels of God.

("sons of God"). ***Job 38:7; "When the morning stars sang together, and all the sons of God shouted for joy?"***

After his fall, he deceived the naïve and innocent man *Hebrew: adam:* אדם to recover his former authority over the Earth, an authority given to man by God at man's creation in day six. ***Genesis 1:28; Matthew 4:8-9; and Luke 4:5-7; Genesis 3:4-5; "And the serpent said unto the woman, Ye shall not surely die: For God doth know that in the day ye eat thereof, then your eyes shall be opened, and ye shall be as gods, knowing good and evil."*** If Satan could get the ones with the dominion over all the Earth (man) to sin, then he would once again control his former dominion, since he was the one who controlled sin, and those who sin. When Adam and Eve sinned against God, they changed their God given exalted nature of righteousness to an abased nature of sin. They relinquished their exalted right to the authority over everything to him who had authority over their new abased nature. He who had the dominion over sin, and all who sin.

The following two verses, from the gospels, prove Satan again controlled the dominion over all the Earth, and over man and his kingdoms, because, if he didn't, this was a hollow promise made to Jesus and no temptation at all.

Matthew 4:8-9; "Again, the devil taketh him up into an exceeding high mountain, and sheweth him all the kingdoms of the world, and the glory of them; And saith unto him, All these things will I give thee, if thou wilt fall down and worship me."

Luke 4:5-7; "And the devil, taking him up into an high mountain, shewed unto him all the kingdoms of the world in a moment of time. And the devil said unto him, All this power will I give thee, and the glory of them: for that is delivered unto me; and to whomsoever I will I give it. If thou therefore wilt worship me, all shall be thine."

There are another three blocks of scripture which reveal more about the heinous and evil nature of the deceitful devious working of this reprobate being. One is in *Job 1:6-12;* and one is in *Job 2:1-7;* and one in *Revelation 12:7-10;*.

Revelation 12:7-10; "And there was war in heaven: Michael and his angels fought against the dragon; and the dragon fought and his angels, And prevailed not; neither was their place found any more in heaven. And the great dragon was cast out, that old serpent, called the Devil, and Satan, which deceiveth the whole world: he was cast out into the earth, and his angels were cast out with him. And I heard a loud voice saying in heaven, Now is come salvation, and strength, and the kingdom of our God, and the power of his Christ: for the accuser of our brethren is cast down, which accused them before our God day and night."

Job 1:6-12; "Now there was a day when the sons of God came to present themselves before the LORD, and Satan came also among them. And the LORD said unto Satan, Whence comest thou? Then Satan answered the LORD, and said, From going to and fro in the earth, and from walking up and down in it. And the LORD said unto Satan, Hast thou considered my servant Job, that there is none like him in the earth, a perfect and an upright man, one that feareth God, and escheweth evil? Then Satan answered the LORD, and said, Doth Job fear God for nought? Hast not thou made an hedge about him, and about his house, and about all that he hath on every side? thou hast blessed the work of his hands, and his substance is increased in the land. But put forth thine hand now, and touch all that he hath, and he will curse thee to thy face. And the LORD said unto Satan, Behold, all that he hath is in thy power; only upon himself put not forth thine hand. So Satan went forth from the presence of the LORD."

Job 2:1-7; "Again there was a day when the sons of God came to present themselves before the LORD, and Satan came also among them to present himself before the LORD. And the LORD said unto Satan, From whence comest thou? And Satan answered the LORD, and said, From going to and fro in the earth, and from walking up and down in it. And the LORD said unto Satan, Hast thou considered my servant Job, that there is none like him in the earth, a perfect and an upright man, one that feareth God, and escheweth evil? and still he holdeth fast his integrity, although thou movedst me against him, to destroy him without cause. And Satan answered the LORD, and said, Skin for skin, yea, all that a man hath will he give for his life. But put forth thine hand now, and touch his bone and his flesh, and he will curse thee to thy face. And the LORD said unto Satan, Behold, he is in thine hand; but save his life. So went Satan forth from the presence of the LORD, and smote Job with sore boils from the sole of his foot unto his crown."

After his wretched fall and until the woebegone War in Heaven, **Revelation 12:7-10;** Satan's frequent and recurrent access to the majestic and august presence of God was as the insidious and critically pointed accuser of the pious and saintly brethren. ***Job 1:6-12; Job 2:1-7;*** and ***Revelation 12:10;*** In fact, he, and all the angels, were required to present themselves before God at regular intervals. ***Job 1:6; Job 2:1***;

After he re-acquired the authority of the Earth from man, he spent his days walking to and fro upon the Earth, and walking up and down in it. ***Job 1:7; and Job 2:2;***

God created him to destroy. ***Isaiah 54:16***; ***"Behold, I have created the smith that bloweth the coals in the fire, and that bringeth forth an instrument for his work; and I have created the waster to destroy."*** This could have a double meaning, first, God created him (Satan) to be a destructive force, a destroyer, which he most assuredly fulfills, or, second, God created him to destroy him, which God will surely do in the lake of fire.

He stood up against the whole nation of Israel and provoked King David to number Israel, to perform a census against the express command of God. ***I Chronicles 21:1;*** ***"And Satan stood up against Israel, and provoked David to number Israel."***

He tried to resist Joshua, the high priest, as Joshua stood before the angel of the LORD, and the LORD rebuked Satan there and compared him to a brand plucked from the fire. ***Zechariah 3:1-2; "And he shewed me Joshua the high priest standing before the angel of the LORD, and Satan standing at his right hand to resist him. And the LORD said unto Satan, The LORD rebuke thee, O Satan; even the LORD that hath chosen Jerusalem rebuke thee: is not this a brand plucked out of the fire?"***

The New Testament

He sinned from the beginning, ***I John 3:8; "He that committeth sin is of the devil; for the devil sinneth from the beginning. For this purpose the Son of God was manifested, that he might destroy the works of the devil."***

His fall could be very near the beginning, when God created the Heaven and the Earth, or much later, as will be shown. Jesus is saying sin was latent in this most exalted of beings from the beginning. Jesus was manifested to destroy his evil and pernicious works, which He accomplished at Calvary.

Jesus beheld him fall from Heaven as lightning, swift and certain, **Luke 10:18;** *"And he said unto them, I beheld Satan as lightning fall from heaven."*

He binds people with crippling infirmity for years. **Luke 13:16;** *"And ought not this woman, being a daughter of Abraham, whom Satan hath bound, lo, these eighteen years, be loosed from this bond on the sabbath day?"*

He tempted Jesus in the wilderness after Jesus spent forty days there praying and fasting. **Matthew 4:1-11; Mark 1:12-13; and Luke 4:1-13;**

Mark 1:12-13; *"And immediately the Spirit driveth him into the wilderness. And he was there in the wilderness forty days, tempted of Satan; and was with the wild beasts; and the angels ministered unto him."*

He is the enemy of man. **Matthew 13:39;** *"The enemy that sowed them is the devil; the harvest is the end of the world; and the reapers are the angels."*

He is a thief. **Luke 8:12; and John 10:10;** *"Those by the way side are they that hear; then cometh the devil, and taketh away the word out of their hearts, lest they should believe and be saved. The thief cometh not, but for to steal, and to kill, and to destroy: I am come that they might have life, and that they might have it more abundantly."*

He was a murderer from the beginning and he never walked in the truth, in fact, there is no truth in him. He is a liar and all lies spout from him, he is the father of lies. **John 8:44;** *"Ye are of your father the devil, and the lusts of your father ye will do. He was a murderer from the* *beginning, and abode not in the truth, because there is no truth in him. When he speaketh a lie, he speaketh of his own: for he is a liar, and the father of it."*

This seems to imply at his fall there was a murder, perhaps of Lucifer himself, the exalted angel, or perhaps he was angry with God and this was likened to murder by Jesus.

He is the prince of this world. *John 12:31; "Now is the judgment of this world: now shall the prince of this world be cast out."* His judgment came at Calvary.

John 14:30; "Hereafter I will not talk much with you: for the prince of this world cometh, and hath nothing in me."

He is the god of this world. *II Corinthians 4:4; "In whom the god of this world hath blinded the minds of them which believe not, lest the light of the glorious gospel of Christ, who is the image of God, should shine unto them."*

He is the prince of the power of the air. He is the spirit of disobedience. *Ephesians 2:2; "Wherein in time past ye walked according to the course of this world, according to the prince of the power of the air, the spirit that now worketh in the children of disobedience."*

He entered into the apostle Judas Iscariot, the betrayer of Jesus to the Jewish elders and the Romans. *Luke 22:3; "Then entered Satan into Judas surnamed Iscariot, being of the number of the twelve".*

John 13:27;"And after the sop Satan entered into him. Then said Jesus unto him, That thou doest, do quickly."

He filled the heart of Ananias and his wife Sapphira to lie to the Holy Spirit. *Acts 5:3; "But Peter said, Ananias, why hath Satan filled thine heart to lie to the*

Holy Ghost, and to keep back part of the price of the land?"

He is transformed into an angel of light. *II Corinthians 11:14; "And no marvel; for Satan himself is transformed into an angel of light."*

He deceives the whole world. He was cast out of Heaven unto the Earth with all his angels, the wicked angels, after the failed War in Heaven. *Revelation 12:9; "And the great dragon was cast out, that old serpent, called the Devil, and Satan, which deceiveth the whole world: he was cast out into the earth, and his angels were cast out with him."*

He sets snares for us. *I Timothy 3:7; "Moreover he must have a good report of them which are without; lest he fall into reproach and the snare of the devil."*

He will throw some of us into prison. *Revelation 2:10; "Fear none of those things which thou shalt suffer: behold, the devil shall cast some of you into prison, that ye may be tried; and ye shall have tribulation ten days: be thou faithful unto death, and I will give thee a crown of life."*

He exercised the power and authority over death, until Jesus, through death, took the power from him. *Hebrews 2:14; "Forasmuch then as the children are partakers of flesh and blood, he also himself likewise took part of the same; that through death he might destroy him that had the power of death, that is, the devil;"*

If you resist him, he will flee from you. *James 4:7; "Submit yourselves therefore to God. Resist the devil, and he will flee from you."*

He walks about like a roaring lion seeking whom he may devour. He is our adversary. He is voracious like a lion. *I Peter 5:8; "Be sober, be vigilant; because your adversary the devil, as a roaring lion, walketh about, seeking whom he may devour:"*

He contended with Michael about the body of Moses. ***Jude 9; "Yet Michael the archangel, when contending with the devil he disputed about the body of Moses, durst not bring against him a railing accusation, but said, The LORD rebuke thee."***

At the War in Heaven, he cast a third of the angels to the Earth. ***Revelation 12:4; "And his tail drew the third part of the stars of heaven, and did cast them to the earth: and the dragon stood before the woman which was ready to be delivered, for to devour her child as soon as it was born."***

He led the wicked angels in a War in Heaven against Michael and the good angels and he lost the war. ***Revelation 12:7-9; "And there was war in heaven: Michael and his angels fought against the dragon; and the dragon fought and his angels, ⁸And prevailed not…"***

When he lost the War in Heaven, he was not allowed in Heaven anymore, neither were his angels. ***Revelation 12:8; "…neither was their place found any more in heaven."***

He deceives the whole world. He and his angels were cast out into the Earth. ***Revelation 12:9; "And the great dragon was cast out, that old serpent, called the Devil, and Satan, which deceiveth the whole world: he was cast out into the earth, and his angels were cast out with him."***

He has great wrath because he knows he has just a short time. ***Revelation 12:12; "Therefore rejoice, ye heavens, and ye that dwell in them. Woe to the inhabiters of the earth and of the sea! for the devil is come down unto you, having great wrath, because he knoweth that he hath but a short time."***

He gives power and great authority to the beast. And when the beast is wounded and is healed, all the world worships the dragon, Satan. ***Revelation 13:2-4;*** *"And the beast which I saw was like unto a leopard, and his feet were as the feet of a bear, and his mouth as the mouth of a lion: and the dragon gave him his power, and his seat, and great authority. And I saw one of his heads as it were wounded to death; and his deadly wound was healed: and all the world wondered after the beast. And they worshipped the dragon which gave power unto the beast: and they worshipped the beast, saying, Who is like unto the beast? who is able to make war with him?"*

He lost his authority over the Earth when Jesus took the keys of hell and of death away from him, at the resurrection. ***Revelation 1:18;*** *"I am he that liveth, and was dead; and, behold, I am alive for evermore, Amen; and have the keys of hell and of death."*

He will be bound with a great chain and imprisoned in the pit for a thousand years after the Battle of Armageddon. He deceives the nations. ***Revelation 20:1-3;*** *"And I saw an angel come down from heaven, having the key of the bottomless pit and a great chain in his hand. And he laid hold on the dragon, that old serpent, which is the Devil, and Satan, and bound him a thousand years, And cast him into the bottomless pit, and shut him up, and set a seal upon him, that he should deceive the nations no more, till the thousand years should be fulfilled: and after that he must be loosed a little season."*

After the thousand years, he will be released from his prison (the bottomless pit) for a little season. He will gather an innumerable army from the four quarters of the Earth to march upon the camp of the saints, and upon the beloved city, Jerusalem. He will be stopped when fire from God comes down out of Heaven. He will then be cast into the lake of fire where he shall be tormented day and night forever and ever, and where the beast and the false prophet are, before the Great White Throne Judgement. ***Revelation 20:7-10; "And when the thousand years are expired, Satan shall be loosed out of his prison, And shall go out to deceive the nations which are in the four quarters of the earth, Gog and Magog, to gather them together to battle: the number of whom is as the sand of the sea. And they went up on the breadth of the earth, and compassed the camp of the saints about, and the beloved city: and fire came down from God out of heaven, and devoured them. And the devil that deceived them was cast into the lake of fire and brimstone, where the beast and the false prophet are, and shall be tormented day and night for ever and ever."***

The Lake of Fire (everlasting fire) was prepared by God for the devil (Satan) and his angels. ***Matthew 25:41;"Then shall he say also unto them on the left hand, Depart from me, ye cursed, into everlasting fire, prepared for the devil and his angels:*** According to scripture, it is uncertain where this lake of fire exists, or will exist when it is filled by the devil and his rebellious and wicked angels.

The Timeline:

Satan was once a bright and shining angel in the times before his ill-fated fall. ***Ezekiel 28:17;*** He held dominion over the Earth and all its creatures. ***Isaiah 14:20;*** He was in an exalted position in the hierarchy of Heaven, and in the service of God. ***Ezekiel 28:14***;

When Satan fell, he lost this dominion. God took it back from him, and destroyed the Old Earth which was once also his dominion. ***Ezekiel 28:16 & 18; Jeremiah 4:23-28;***

Then long ages passed and God repented of destroying the Old Earth. *Jeremiah 4:27;* Then God reconstructed the destroyed Old Earth into the one man inhabits today, *Genesis 1:2-31;* and God created man. *Genesis 1:26-28;* He gave man dominion, *Genesis 1:26;* the dominion once belonging to Satan. *Ezekiel 28:14;* Man was given dominion and authority over all the elements of the Earth, *Genesis 1:26;* elements which God reestablished at the Reconstruction. *Genesis One;* Satan wanted this dominion back and when God placed Adam and Eve in the Garden of Eden, *Genesis 2:15;* Satan saw his chance. *Genesis 3:1;*

Satan didn't have a chance until God gave man a commandment forbidding some behavior. *Genesis 2:16-17;* The minute God gave Adam and Eve the commandment not to eat of the Tree of the Knowledge of Good and Evil, Satan formulated his plan. *Genesis 3:1;*

He tempted man with a lie, *Genesis 3:1-6;* causing him to partake of Satan's own nature, the sin nature, *Ezekiel 28:15-16;* instead of the nature of the LORD God, righteousness, *Revelation 4:8;* with whom he enjoyed fellowship up until then. *Genesis 3:8;* He lied to the man, *Genesis 3:1;* and the man believed him. *Genesis 3:6;* Once Satan obtained the dominion again, he set about enslaving men as their new master. *John 14:30;*

However, God made provision for man's release from Satan's hold. He gave the promise of a redeemer. *Genesis 3:15; Genesis 49:10;*

From the moment the promise was made, Satan set about to destroy the means by which this redemption promise could be fulfilled. He was certain this redeemer must of necessity, by the nature of the very promise, be born a man, through a righteous (chosen) lineage. When God accepted Abel's humble sacrifice *Genesis 4:4*; and denied Cain's *Genesis 4:5*; Satan took this as a sign God was identifying the line through which the redeemer would come and Satan tempted Cain to rise up and violently kill Abel. *Genesis 4:8;* Cain killed Abel from the malevolent evil residing within himself, in his own heart, but Satan was surely behind it, to accomplish his impious desires. *Genesis 4:6-7;* He thought his irascible problem was satisfactorily solved, but he was sorely mistaken, God merely reestablished the chosen line through Seth, Abel's younger brother and suitably acceptable replacement as the instrument of serendipity for man's future. *Genesis 4:25*; If God hadn't driven Cain out from the place where Adam and Eve were dwelling east of the Garden of Eden, and sent him away to the land of Nod, Cain would most likely have killed Seth also, because Cain proved he was susceptible to sin and his nature was to succumb to the sin lying in wait at his door. *Genesis 4:7;* Even though Satan surely tempted Cain to kill Abel, it was within Cain's violently inclined heart to kill Abel in the first place.

  It must be noted, before Jesus' resurrection, Satan had no power to directly kill a man, God restricted him. When he presented himself before the LORD *Job 1; and Job 2*; he needed permission to even touch Job's body, but the LORD restrained him from killing Job. And after Jesus came, Satan no longer controlled dominion over death because Jesus took away the keys of Hell and of Death. *Revelation 1:18;* Now Jesus has the authority over death and Satan has no power to directly kill a man.

  When a man kills another man, it is from the evil resident in his own unrepentant heart. Cain killed Abel because of the evil resident within. It served Satan's evil purposes, but it came from Cain. That's why God told Cain *"...sin lieth at his door..."*, and he could control it, or it would control him. ***Genesis 4:7;*** This is a lesson for all men, sin lies at the door, however, a believer can control it.

The rest of the history of the Jewish people down to the time of Jesus' timely and favorable birth, shows numerous signs of Satan trying to eliminate the line through which the redeemer would come. His role was still one of master of the human race, as evidenced by the fact he controlled the underworld, Hades and Paradise. This is the significance of the keys of Hell and Death. ***Matthew 16:19; Revelation 1:18;*** The righteous dead rested in Paradise and the wicked dead in Hades. ***Luke 16:19-31;***

When Jesus went back to Heaven at the Ascension, ***Acts 1:9;*** he took the captives with him, ***Ephesians 4:8-10;*** emptying the underworld, both Hades and Paradise.

When Jesus came, the nature of Satan's business changed. He was no longer occupied with destroying the line through which the redeemer would come, since the redeemer was already there. The temptation in the wilderness showed this to Satan in glaring detail. ***Matthew 4:1-11; Mark 1:12-13; Luke 4:1-13;***

Satan shifted his plan, and boldly marched his angels on Heaven, and fought the War in Heaven against Michael and the good angels at this time, the time when Jesus was upon the Earth. ***Revelation 12:7;*** See; the author's book; **The Great Tribulation and the War in Heaven.**

This is the manifestation of the beast, who at this time is called the Great Dragon, that old devil, and Satan, ***Revelation 12:3;*** who is revealed as a beast in three other places also. ***Daniel 7:7; Revelation 13:1; Revelation 17:3***; Each representation of this beast in scripture is a manifestation of Satan at four different times in man's history, as was shown in the author's work **"The Apotheosis of the Apocalypse".**

When he lost this War in Heaven, he was no longer allowed access to Heaven, *Revelation 12:8;* he no longer played the role of the *"accuser of the brethren". Revelation 12:10;* He was cast to the Earth, *Revelation 12:9;* and this time, he was not in the role of the one with dominion, Jesus was now the one with dominion. *Revelation 1:18;* Satan now played the role of the spoiler, *John 10:10;* the tempter, *James 1:14;* the deceiver, *Revelation 20:3;* the liar, *John 8:44;* the evil one and the destroyer. *John 10:10*;

His role now was to deceive as many men as he could to keep them from the fulfilled promise accomplished upon the Earth. And this role he would continue until the Second Coming of Jesus to the Earth, when his role will change again. This role of destroyer is described as another beast. *Revelation Thurteen*;

At the Second Coming, Satan will gather the armies of the Earth to wage war against the returning redeemer and the armies which were in Heaven *Revelation 19:14;* at the Battle of Armageddon. *Revelation 16:14-16;* His role changes again at this time. Why he feels he can defeat God with a few armies of men is beyond belief, since he was unable to overthrow the heavenly army with his angels before. He nonetheless gathers these armies together for one sole purpose, the Battle of Armageddon, and he loses this battle also. *Revelation 19:19-20;* This is described as the beast *Revelation Seventeen;* upon whom Mystery Babylon rides.

At the end of the terrible Battle of Armageddon, a battle more destructive and horrible than any other battle ever fought upon Earth, Satan will be taken and bound in the bottomless pit for one thousand years, *Revelation 20:2-3;* and Jesus will reign upon the Earth during this halcyon period. *Revelation 20:4;* The pernicious influence of Satan will be entirely absent among men, and Jesus will rule all men with a rod of iron. **Revelation 12:5; Revelation 19:15;**

At the end of this one thousand years, Satan will be released from the pit **Revelation 20:7;** and will go throughout the Earth and gather an army numbering as the sand of the sea. **Revelation 20:8;** It seems unusually extraordinary Satan would be able to gather such a large army from among men, because they just lived through a one thousand year period of peace and tranquility. Remember, Jesus rules this one thousand years with a strict and rigid rod of iron, and Satan's manipulating influence was not felt during this entire time. When Satan is released, he goes throughout the Earth spewing the same old lies he has always used, and deceives men, turning them away from God. He tells them they don't have to do what Jesus says, and they believe him. This is the beast at the end of the age. **Daniel 7; Revelation 20**;

Know this, Satan is not a stupid being, if he didn't think he could win, he would not partake of an exercise in futility. He didn't win the War in Heaven with the angels, and be sure, over the millennia he has thought long and hard about why he lost the War in Heaven. He has thought it through and he thinks he has a plan to win at Armageddon, using men as his minions instead of angels.

When this attempt also fails, he will sit in the pit for one thousand years, analyzing his failure, and coming up with, what will seem to him, a sure-fire plot to win when he is finally released.

The book is already written, and the end already assured, but this means nothing to him. He is trying to beat God, and it is God who has written the book, and assured his end. If he beats God, he becomes God, and he can then write his own book, with its own end.

At least this is how he sees it, through his own perverted and reprobate mind. He freely partook of sin, and his iniquity consumed him. *Ezekiel 28:18;*

The Fall of Lucifer

One thing about Satan, at the beginning, *Job 38:7;* he was not yet fallen. At what time he fell, the scriptures do not elucidate. There are, however, other indicative clues leading to this truth.

When he tempted Adam and Eve in the Garden of Eden, he was already a fallen being. Sometime between the creation of the Earth, and the placing of Adam in the Garden of Eden, Satan fell.

The Earth, according to scientist's best estimate, is about 4.5 billion years old. Sometime between 4.5 billion years ago, and about six thousand years ago, when Adam and Eve were in the Garden of Eden, Satan fell.

Since we know from scripture, at Satan's fall, God destroyed the Old Earth completely, we need only look at the record in the fossils of the Earth for a comparable destruction to determine the timing of his fall. If an ascertainment of any time in the history of the Earth when a level of destruction this absolute occurred, the time when Satan fell can, maybe, be pinpointed.

There are two separate events which qualify, when comets (or asteroids or perhaps another planetary body) struck the Earth causing tremendous global destruction. Either of these could have been representing the vision Jeremiah experienced *Jeremiah 4:23-28;* and both will be explained although the second one is preferable as will be shown.

The first was when the dinosaurs went extinct at 65 million years ago, when an asteroid of about 9.6 kilometers

(6 miles) in diameter struck the Earth just off the Yucatan Peninsula. During this catastrophic event, most of the flora and the fauna, with some estimates as high as seventy five percent (75%) went extinct, and though this was probably not the event wrought by the LORD, the cause for the destruction of the Old Earth leaving it in a state of chaos, **Genesis 1:2;** it does show a mechanism (an asteroid) by which extinction on a major scale may be accomplished.

The other event is one causing the near complete extinction (ninety nine percent (99%) of the living thriving species perished at this disastrous epochal event) of the flora and fauna in the mid-Permian Period, about 240 million years ago. This event also qualifies as the event of total destruction which God brought about because of his fierce anger, **Jeremiah 4:26; Genesis 1:2;** and therefore may allow us to place the time of the fall of Satan.

However, there were other and greater mass extinctions before the Great Dying of the mid-Permian period. Those extinctions were of lower forms of life, lower than man. Life had not progressed to the level of intricacy necessary to bring forth man at those earlier times. Scientists do not consider man to have emerged before the K-T extinction, however, with the time frames involved, it is possible the Earth could have been populated with men, fruitful places and cities as early as before the mid-Permian, whose traces have long since completely vanished. This is why we will confine our God induced mass extinction to the one 240 million years ago, and therefore pinpoint the fall of Satan to this time.

This work will examine all the major extinction events in more detail later.

The Reconstruction

Since this thesis has established the Gap, and the vast ages of scientific creationism comes sliding correctly into place, it is necessary to ask, What are the descriptions given in **Genesis One;** describing, if not the individual acts of creation?

This narrative in **Genesis One;** describes the Reconstruction of the Earth by God, made necessary by the total devastation previously wrought by God, and not the creation, an act accomplished by God, *"In the beginning...".* **Genesis 1:1;** This Reconstruction could have been accomplished by God at any time after He destroyed the Old Earth, which this study conjectured occurred during the mid-Permian Great Dying event (240 million years ago). This Reconstruction could have been accomplished as early as 239 million years ago, leading to the predominant rise of the dinosaurs, and their sudden extinction at the K-T boundary (65 million years ago), which led to the ascendancy (creation) of man. Scientists are in general agreement early man arose after the extinction of the dinosaurs, however, there is evidence in a river-bottom in Texas of human footprints existing directly adjacent to those of a dinosaur, meaning they both crossed the still hardening surface of the sands there at the same time, or they co-existed. This evidence points to a much earlier origination of day six man than scientists are willing to consider. This would mean their emergence corresponds to this time in prehistory, at least 65 million years ago, and this gives a time anchor to day six of the Reconstruction.

The Reconstruction of Genesis One;

This thesis will now undertake to describe the Reconstruction of the Earth, from a state of devastation, to a state of habitability, as described in **Genesis One;**. God caused the destruction, and He accomplished the Reconstruction.

An analysis of the various acts being described in **Genesis One;** clearly allows one to perceive the Reconstruction, and in seeing the Reconstruction, get a better understanding of the extent of the destruction occasioned by Satan's fall.

The Face of the Deep and the Face of the Waters

Genesis 1:2; "And the earth was without form, and void; and darkness was upon the face of the deep. And the Spirit of God moved upon the face of the waters."

As this thesis previously discussed the "without form" and "void" aspects of *verse two*; in another section, it is now necessary to examine the remainder of the verse.

There was darkness upon the face of the deep. *Hebrew: tehom:* תְּהוֹם *the abyss, a great expanse of water, a subterranean sea* Later, in *verse 5*; God identified the darkness as night, and night was upon the face of the deep. This means the side of the Earth which contained the deep, was continually on the dark side of the Earth. This could indicate the Earth was rotating extremely slowly, just like Earth's Moon rotates today, always keeping one side toward the Sun, or it was not rotating at all, meaning it was acting more like a comet in its revolution around the Sun.

The scripture describes the deep as containing a face. It is necessary for clarification, to identify this "deep", to fully understand what God was accomplishing here. In our section on the Agent of Mass Destruction responsible for this carnage, is a proposition to identify the current Pacific Ocean as the sheared off side of the Earth (the deep), and these verses explain this sheared off side of the Earth as occupying the dark side.

The scriptures also says the waters exposed a face, and it is easily assumed the reference is to the face of the ocean, the surface. This further identifies the "deep" as a place of water, identifying both as having a "face". It also says in the next verse, the water covered the whole planet, but the deep was only on one side, this is why it also identified the water as showing a face, the surface, distinguishing it from the deep, which besides exhibiting a face, a surface, was also shrouded in darkness.

Light

Genesis 1:3-4; "And God said, Let there be light: and there was light. And God saw the light, that it was good: and God divided the light from the darkness. And God called the light Day, and the darkness he called Night. And the evening and the morning were the first day."

Then light came into being. If a celestial body was wrapped in darkness, not rotating, the continuation of its previously retarded rotation, would again cause the light to be seen. The light would be diffuse, without any apparent discernable source, until a steady rate of rotation was again established, and perhaps a steady orbit as well.

The Earth, after the collision with the rogue planet, was acting more like an immense comet, than a stable planet. It was in an unpredictable and erratic path through the Solar System, out beyond the orbit of Mars, out to the asteroid belt, or beyond at aphelion, and maybe in as far as the orbit of Venus during perihelion, maybe even striking

Venus and Mars during its erratic flight. The Earth was also partly responsible for escorting many of the asteroids, the remnants from the primordial collision, out to their present orbit. The near Earth orbiting asteroids threatening Earth today, are the ones remaining from the Earth/Agent encounter, the fragments not escorted to the asteroid belt by the marauder on its outward trajectory or by the Earth in its early and severely elongated orbit at aphelion. After the Earth again began a regular rotation it acquired its current stable orbit, between Venus and Mars, and settled down.

What could possibly cause the Earth to again begin its regular rotation?

The Moon struck the Earth, at this time, at a specific point on the floor of the South Pacific Ocean near the Mariana's Trench just off the coast of the Mariana Islands, where the rare and uncommon silica found on the Moon are also found on the Earth, the only place those scarce and unusual silica are found on Earth. These rare silica are abundant on the Moon.

This micro collision, by the Moon, caused the immobile stationary Earth to begin spinning at a tremendous rate of speed. This would account for the diffused light, the Earth was spinning too fast to make out a source for the light. It would also account for the separation of the waters the next day. The source of the light, the Sun, would appear again on day four.

The Firmament

Genesis 1:6-8; "And God said, Let there be a firmament in the midst of the waters, and let it divide the waters from the waters. And God made the firmament, and divided the waters which were under the firmament from the waters which were above the firmament: and it was so. And God called the firmament Heaven. And the evening and the morning were the second day."

There is much valuable and significant information in this verse answering many questions concerning the early days of man on Earth. God separated the waters from the waters, and placed a firmament in their midst. He divided the waters under the firmament from the waters above the firmament. He called the firmament Heaven, and in *verse 20;* He positively identified the firmament as the atmosphere, the sky where the birds fly above the Earth.

He divided the waters under the atmosphere (the water covering the whole Earth), from the waters above the atmosphere. This is telling us God placed some of the water covering the Earth in *verse 2;* as an enormous cosmic bubble around the Earth. "...*the waters above the firmament...*" This would perhaps occur if the centrifugal force of the newly started rotation, was of sufficient strength to throw the water into low Earth orbit.

This celestial bubble of water would also account for three other issues in *The Book of Genesis;*

First, the ages of the patriarchs. If the patriarchs were protected from the harmful damaging rays of the Sun, by a protective layer of shielding water surrounding the whole globe, they very probably could have lived to the ages of longevity to which they are each ascribed.

Second, the windows of Heaven, *Genesis 7:11;* which were opened up in the days of Noah, could be accounted for by this bubble of water around the whole Earth. This would also account for the absence of the bubble today. Another deleterious encounter with some of the vagrant wandering debris of this first tremendous collision, could have broken the fragile bubble and sent all the water cascading back down to Earth, to flood the entire world in the days of Noah.

Third, this water above and surrounding the Earth would produce a dew or a mist to continually water the Earth. *Genesis 2:6;* It is possible there was no rain during the time this bubble encircled the Earth because of the constantly humid condition of the atmosphere.

This ubiquitous and beneficial mist was present in the morning and watered the fruitful places, nourishing the crops.

The Sea, the Land and Plant Life (Flora)

Genesis 1:9-13; "And God said, Let the waters under the heaven be gathered together unto one place, and let the dry land appear: and it was so. And God called the dry land Earth; and the gathering together of the waters called he Seas: and God saw that it was good. And God said, Let the earth bring forth grass, the herb yielding seed, and the fruit tree yielding fruit after his kind, whose seed is in itself, upon the earth; and it was so. And the earth brought forth grass, and herb yielding seed after his kind, and the tree yielding fruit, whose seed was in itself, after his kind: and God saw that it was good. And the evening and the morning were the third day."

The waters gathered into one place, not the seven ocean basins, just one ocean basin, draining off the land, filling the deep, and allowing the primeval continent of Gondwanaland to appear. The seven seas did not appear until later, when continental drift and plate tectonics allowed the water to fill the gaps between the separating land masses. Originally, at the beginning of the

Reconstruction, there was only one ocean, on one side of the Earth, while the continent was on the other side.

This is also conjectured in scientific circles. Scientists call the ocean Panthalassa and they call the continent Gondwanaland.

The next step for the continents was for grass and plants and trees to take root and begin to grow. In this Reconstruction event, God placed the means of reproduction within each plant and animal to self-propagate each separate and distinct species.

The Sun, the Moon and the Stars

Genesis 1:14-19; "And God said, Let there be lights in the firmament of the heaven to divide the day from the night; and let them be for signs, and for seasons, and for days, and years: And let them be for lights in the firmament of the heaven to give light upon the earth: and it was so. And God made two great lights; the greater light to rule the day, and the lesser light to rule the night: he made the stars also. And God set them in the firmament of the heavens to give light upon the earth. And to rule over the day and over the night, and to divide the light from the darkness: and God saw that it was good. And the evening and the morning were the fourth day."

It doesn't say God created the Sun and the Moon and the stars at this time, it merely says He set them in the firmament.

Now is the time for an analysis of the word "set" (*Hebrew: nathan:* נתן *to put, to fasten, restore)* to see what it reveals. When it says God set the Sun, the Moon and the stars in the firmament of the Heaven, contemporary thought interprets it to mean He set (placed) them in the Heaven. It could also mean He set (stabilized) them in the Heaven.

What this could and probably does mean, in context with the Reconstruction, is He was really stabilizing the Earth in its daily rotation and annual revolution, thereby giving the impression of stabilizing the Sun, the Moon and the stars from the Earthly perspective.

Marine Life and Fowls (Fauna)

***Genesis 1:20-23;** "And God said, Let the waters bring forth abundantly the moving creature that hath life, and fowl that may fly above the open firmament of heaven. And God created the great whales, and every living creature that moveth, which the waters brought forth abundantly, after their kind, and every winged fowl after his kind: and God saw that it was good. And God blessed them, saying, Be fruitful, and multiply, and fill the waters in the seas, and let fowl multiply in the earth. And the evening and the morning were the fifth day."*

The fifth day saw the propagation of the fowl of the air, and the creation of the great whales.

This is one of the three elements in **Genesis One;** which God created , *Hebrew: bara:* בָּרָא, *to make something new, something not there before, from materials not present before,* the first being the Heavens and the Earth in **verse 1;**, and the other being man in **verse 27;.** These are the only three times God actually created something in **Genesis One;.**

Land Animals (Fauna)

***Genesis 1:24-25;** "And God said, Let the earth bring forth the living creatures after his kind, cattle, and creeping thing, and beast of the earth after his kind: and it was so. And God made the beast of the earth after his* *kind, and cattle after their kind, and every thing that creepeth upon the earth after his kind: and God saw that it was good."*

The cattle and all the moving and creeping creatures were made next. They were invested with the same ability to propagate their species being inherent in their structure. God was now finished setting the scene for man, the object of all this activity.

The Creation of Man

Genesis 1:26-28; "And God said, Let us make man in our image, after our likeness: and let them have dominion over the fish of the sea, and over the fowl of the air, and over the cattle, and over all the earth, and over every creeping thing that creepeth over the earth. So God created man in his own image, in the image of God created he him; male and female created he them. And God blessed them, and God said unto them, Be fruitful, and multiply, and replenish the earth, and subdue it: and have dominion over the fish of the sea, and over the fowl of the air, and over every living thing that moveth upon the earth."

He made man in His own image. What image is this? Does God look like us, or is there another meaning to this verse. *John 4:24;* says, *"God is a Spirit..."* and this is the nature and image of God. It is the image God imparted to man when He *"...breathed into his nostrils the breath of life"* spirit, *"...and man became a living soul." Genesis 2:7;*

This study will revisit the creation of man again shortly.

Man's Diet and All Other Creatures Food Sources

Genesis 1:29-31; "And God said, Behold, I have given you every herb bearing seed, which is upon the face of all the earth, and every tree, in the which is the fruit of a tree yielding seed; to you it shall be for meat. And to every beast of the earth, and to every fowl of the air, and to every thing that creepeth upon the earth, wherein there is life, I have given every green herb for meat: and it was so. And God saw every thing that he had made, and, behold, it was very good. And the evening and the morning were the sixth day."

Man was originally a strict vegetarian. It wasn't until after Noah came back to solid ground and debarked from the Ark before God allowed man to eat meat. ***Genesis 9:3;***

The Day of Rest (Sabbath)

Genesis 2:1-3"Thus the heavens and the earth were finished, and all the host of them. And on the seventh day God ended his work which he had made; and he rested on the seventh day from all his work which he had made. And God blessed the seventh day and sanctified it: because that in it he had rested from all his work which God created and made."

In these verses, it emphasizes three times these are the acts which God made, *(Hebrew: asah:* עָשָׂה: *to make something new, but from existing materials),* and the verse ends with "*...because that in it he had rested from all his work which God created and made."* distinguishing the acts of "creating" and "making". In ***Genesis One;*** there is another word used also in the acts performed translated as "formed". *Hebrew: yatsar:* יָצַר *to fashion something which was previously present, from existing materials*

Genesis 2:4"These are the generations of the heavens and of the earth when they were created, in the day that the LORD God made the earth and the heavens."

This verse confirms the six day Reconstruction was not the original creation, since this verse asserts, *"...in the day that the LORD God made the earth and the heavens."* God made the Earth and the Heavens in one day, *"In the beginning..." Genesis 1:1;*

What is Man?

In the six days of Reconstruction, **Genesis One;** God performs a different act each day, culminating with the creation of man on the sixth day, and a day of rest on the seventh.

The most common interpretation, held by biblical scholars, of the acts God is performing in **Genesis One;** is it was the original creation. As has already been shown, it was not the original creation, but, rather, a Reconstruction of the Earth to a habitable state from an uninhabitable one.

The Hebrew word for create is *Hebrew: bara:* אָבָּרְ, which means to make something new from material not existing prior to the act of creation, like the original creation, making something (the Universe) out of nothing. **Hebrews 11:3;** This is the essence of the meaning of the word *Hebrew: bara:* בָּרָא. as stated, this word was only used three times in **Genesis One;** meaning most of the acts of God in the Reconstruction were not a creation.

It is also commonly believed the man which God created in **Genesis One**; on the sixth day, was Adam and Eve, since it claims God created them male and female. Is this a valid assumption?

It is time to thoroughly scrutinize the sixth day creation of man.

The Creation of Man

Genesis 1:26; "And God said, Let us make man" *(on the sixth day, God created man)* ***"in our image, after our likeness:"*** In the image and likeness of God created He him. The likeness of God is "spirit". ***John 4:24.*** The man created in day six was also a spirit being. ***"...and let them have dominion over the fish of the sea, and over the fowl of the air, and over the cattle, and over all the earth, and over every creeping thing that creepeth upon the earth."***
He gave them dominion over:
1. The fish of the sea {fisher}.
2. The fowl of the air {trapper,}.
3. The cattle {domesticator}.
4. All the Earth {gatherer}.
5. Every living thing that moves upon the Earth {hunter}.

Genesis 1:27-28; "So God created man in his own image, in the image of God created he him; male and female created he them." He created *Hebrew: bara:* בָּרָא man as male and female. ***"And God blessed them, and God said unto them, Be fruitful, and multiply, and replenish the earth,"*** In order to replenish something, it must first be filled, then emptied, then replenished. ***"...and subdue it: and have dominion over the fish of the sea, and over the fowl of the air, and over every living thing that moveth upon the earth."***

Genesis 1:31; "And God saw every thing that he had made, and, behold, it was very good." God proclaimed every single act as 'good' after the completion of each act.

"And the evening and the morning were the sixth day." These were single days, because it says the evening and the morning were the sixth day. Evening and morning makes a full day, not a thousand or more years, as some surmise.

The Creation of Adam

Now comes a look at the creation of Adam.

Genesis 2:4-5; "These are the generations of the heavens and of the earth when they were created, in the day that the LORD God made the earth and the heavens," This emphasizes the Earth and the Heavens were created in a day, not seven days. ***"...and every plant of the field before it was in the earth, and every herb of the field before it grew: for the LORD God had not caused it to rain upon the earth, and there was not a man to till the ground."***

The man created in day six as male and female, were given dominion as fishers, trappers, domesticators, hunters and gatherers, but not as farmers. God tells us there was not a man to till the ground, there was not a farmer.

Genesis 2:6-7; "But there went up a mist from the earth, and watered the whole face of the ground. And the LORD God formed man of the dust of the ground, and breathed into his nostrils the breath of life; and man became a living soul."

God formed Adam *Hebrew: yatsar:* יָצַר *formed, to make something new, from existing material;* as opposed to create; *Hebrew: bara:* בָּרָא *to make something new from material not previously present, or from nothing* which is what God did concerning man in day six. In the description of the forming of Adam, the word create *Hebrew: bara:* בָּרָא is not used, as it is in the day six creation of man. He formed the man of the dust of the ground, gave him the breath of life and he became a living soul.

Genesis 2:8; "And the LORD God planted a garden eastward in Eden; and there he put the man whom he had formed."

Then God planted the Garden of Eden, after He formed the man. This planting of the garden was a process which takes time, at least one growing season, probably many more.

Genesis 2:15; "And the LORD God took the man, and put him into the garden of Eden to dress it and to keep it."

He put the man, Adam, which He made before He planted the garden, into the garden to keep it and to dress it. God was letting the man function as a farmer in the Garden of Eden, keeping the plants of the garden, dressing them, arranging them and nurturing them.

Genesis 2:18; "And the LORD God said, It is not good..." This is the first thing, in all the works of Reconstruction, which God proclaimed not good. Every day, after He performed an act, God would look at His work and see it was very good, but this time, God saw it was not good. ***"...that the man should be alone; I will make him an help meet for him."*** God then noticed the man was alone, and needed a help meet. This must have taken quite a while, because loneliness is a process, and takes time. The man was probably in the garden for some time before God noticed his need for a help meet.

Genesis 2:19; "And out of the ground the LORD God formed..." *Hebrew: yatsar:* יָצַר *formed "...every beast of the field, and every fowl of the air; and brought them unto Adam to see what he would call them: and whatsoever Adam called every living creature that was the name thereof."*

This is the first time the man is called Adam, as a proper name. Adam is a Hebrew word meaning 'ruddy man, or red faced man'. It is the same word used for man in the sixth day creation of man, hence the confusion about the creation of man being equated with the creation of Adam and Eve.

God responded to the abject loneliness of Adam by forming the animals out of the ground and allowing Adam to name them. Again this is a process requiring quite some time, a huge undertaking. It must be understood here Adam was naming the animals. He was not naming their genera and classification as scientists do today, he was naming them as a person would name a dog or cat. When God recognized the man's dismal loneliness, His solution was to occupy the man's attention with the time consuming task of naming the animals to alleviate his loneliness. God knew this would not suffice, since He said He would make for him an ameliorative help meet. This naming of the animals was merely a temporary stopgap measure.

Genesis 2:20; "And Adam gave names to all cattle, and to the fowl of the air, and to every beast of the field;" Again, this process must have taken an enormous amount of time, considering the wide variety of the animal kingdom, ***"...but for Adam there was not found an help meet for him."*** And yet God saw Adam was still lonely.

Genesis 2:21; "And the LORD God caused a deep sleep to fall upon Adam, and he slept: and he took one of his ribs, and closed up the flesh instead thereof; And the rib, which the LORD God had taken from man, made he a woman, and brought her unto the man."

Then God made the woman, after all these other events. This was a series of events which must have taken some time indeed. Therefore the making of Eve was not on the same day as the making of Adam, and therefore was not the same as the creation of man on the sixth day of the

Reconstruction since in the creation of man on day six, man was made male and female. ***Genesis 1:27;***

The creation *Hebrew: bara:* בָּרָא of man, male and female, was in one day, on day six, and the making *Hebrew: yatsar:* יָצַר of Adam and Eve, at a later time, was not on the same day.

Cain and His Line

This suitably and conveniently explains where Cain acquired his wife, from the inhabitants of the Earth, the men created on the sixth day. This alleviates the problem of having to conjecture other female children from Adam and Eve, for which there is little to no proof, and from falsely accusing the early inhabitants of the sin of incest, which God expressly forbids. ***Deuteronomy 27:22;***

Why Cain's Sacrifice was Unacceptable

Genesis 3:17-19; *"And unto Adam he said, Because thou hast hearkened unto the voice of thy wife,"* God pronounced Adam's sin as twofold, first he harkened unto the voice of his wife over the voice of God, and second he ate of the Tree of the Knowledge of Good and Evil, a thing which God conclusively forbade in a commandment saying, you shall not eat of it. *"…and hast eaten of the tree, of which I commanded thee, saying, Thou shalt not eat of it:"* Then God pronounces the punishment for the transgression of Adam. *"…cursed is the ground for thy sake;"* God cursed the ground as the man's punishment, Then God says, in a sense, this was a blessing, it was for his (the man's) sake.

His pronouncement to Cain ***Genesis 4:11-12***; was harsher and revealed the ground would never again yield fruit to him for his sin, even if he labored, and this could easily have been the pronouncement against Adam too, if God had wanted to, but for their sake (Adam and Eve's), God pronounced the ground would only yield fruit to them after hard labor, but at least it would yield to them,*"...in sorrow shalt thou eat of it all the days of thy life;"* God said man would eat of the fruit of the ground in sorrow the remainder of his life, and man has, of necessity, labored hard to derive sustenance from the ground ever since. *"...Thorns also and thistles shall it bring forth to thee; and thou shalt eat the herb of the field;"* At least God here allowed man the luxury of eating the fruit of the Earth with herbs and spices to garnish and liven up the food in preparation. *"...In the sweat of thy face shalt thou eat bread,"* God said it would be hard work to gather grain and work the fruit of the ground into edible form. When man resided in the Garden of Eden, everything grew with ease and flourish and was in an edible form immediately for Adam, now he would have to work to make it so. *"...till thou return unto the ground;"* And this would be the case all the days of man's life until he would return to the ground from which he was made, for he was made of dust and unto dust he would return. It is possible man was an immortal being until this pronouncement, making man a mortal, subject to death. *"...for out of it wast thou taken: for dust thou art, and unto dust shalt thou return."* Additionally, God pronounced man's transient ephemeral mortality in this verse.

Original Sin

When Adam and Eve sinned, they were directly cast out of the Garden of Eden. The position of familiar splendid honor they formerly held with the LORD God was now lost, the irreproachable blameless innocence they once knew was now gone forever, for them and for their descendants. They were placed on a par with the ordinary inhabitants of the Earth, the people who were created on the sixth day.

The sin Adam committed was the first sin of the new man on the Earth, and it immediately and conclusively accomplished two stipulations.

First, it separated man from God, because God is holy, and cannot partake of sin.

Second, it passed on a sin nature to all his descendants.

Some scholars teach the sin of Adam (the original sin) is man's indomitable inheritance from Adam and every man at his birth is weightily and profoundly burdened with this woeful legacy. Man needed release from the original sin in order to be able to once again approach God. Our inheritance from Adam was a sin nature, a disposition to sin, the capacity to sin, which was not part of Adam's original makeup. By choosing to sin in the Garden of Eden, Adam passed the nature of sin to all his pitiable progeny, and his unfortunate issue, all mankind, has this inherent capacity within, as part of his nature. Sin carries with it this inherent miserable defect. The individual sins committed in men's lives are a result of this nature inherited from Adam, the inclination to sin when tempted, and man is tempted when drawn away by lusts. ***James 1:14;***

Some scholars teach Jesus was born without this sin nature, being born of a virgin, but the virgin who bare Him was descended from Adam and therefore was a partaker of the same sin nature as the rest of mankind, who are descended from Adam. Therefore she passed it on to her son and Jesus was subject to the sin nature also.

To combat this inconvenient truth, some of these biblical scholars, mostly Roman Catholic in belief, teach Mary too was born as the Immaculate Conception, without sin, as they say, to place Jesus one more generation further removed from Adam's sin nature (original sin). For some inexplicable reason, they see it as a compelling and necessary component of Jesus' character not to be born like the rest of men, with the sin nature, and this somehow explains why He was able to live His life without sinning. It is a statement in itself to say He was born with the same sin nature as the rest of man and paved the way and set the example of man's ability to live a life free of sin.

Jesus was able to live His life without sin because He operated as a person filled with the Holy Spirit and was therefore able to abstain from sin, and this makes His life, and the sacrifice He offered all the more important and relevant to all men, As a man, and as a descendant of Adam, He was able to overcome the influence of sin on man, and thereby redeem man to the Father, to the same position, through Adam, held with Him in the Garden of Eden, one of fellowship. If Jesus did not partake of the sin nature as the rest of humankind does, then He wasn't ever truly tempted by sin, which He was all through His life, yet He did not sin. *Hebrews 4:15;*

The people who were created on the sixth day inhabited the Earth for perhaps thousands of years, *or even millions) an unassuming humble habitation not as exalted as the new creature, Adam. Their simple habitation did not necessarily include a close intimate cherished fellowship with God, yet Adam did have this consummate privilege. The former inhabitants, before Adam, led a simple life, a rudimentary fundamental existence, a hunter/gatherer existence, although it did include an awareness of God, and an afterlife, as explicitly evidenced by the ritual of the burial of their dead. They also lived in cities, and to form the first preliminary traces of what today is called civilization, as evidenced by the early basic settlements in Catal Huyuk, Gobekli Tepi, and perhaps Cappadocia in Anatolia, modern day Turkey, and some of the other stone-age settlements in Eurasia, especially in former Yugoslavia and even into Mesopotamia and Africa.

Adam was created to fill this higher position, an exalted position, which included fellowship with God in this idyllic picturesque serene land, the Garden of Eden. He was to be the tiller of the ground. Then Adam lost his position as "tiller of the ground". Now, tilling the ground would be a backbreaking arduous chore to merely sustain the delicate feeble life to which he now tenaciously held, which would now require a hard day's work to reap benefit.

Cain and Abel

Genesis 4:1-7; "And Adam knew Eve his wife; and she conceived, and bare Cain, and said, I have gotten a man from the LORD. And she again bare his brother Abel. And Abel was a keeper of sheep, but Cain was a tiller of the ground. And in process of time it came to pass, that Cain brought of the fruit of the ground an offering unto the LORD. And Abel, he also brought of the firstlings of his flock and of the fat thereof. And the LORD had respect unto Abel and to his offering: But unto Cain and to his offering he had not respect. And Cain was very wroth, and his countenance fell.
And the LORD said unto Cain, Why art thou wroth? and why is thy countenance fallen? If thou doest well, shalt thou not be accepted? and if thou doest not well, sin lieth at the door. And unto thee shall be his desire, and thou shalt rule over him."

Abel realized his rightful place with the LORD, as a keeper of animals, and his obliged sacrifice was acceptable to God because of this realization, because Abel's sacrifice was blood. God was already establishing the covenant of blood atonement for our sin nature even this early in man's relationship with Himself. He was already setting the stage for the sacrifice of His Son, the sacrifice to take place nearly four thousand years hence, when God's Son Jesus Christ was nailed to the cross, died, was buried and rose again from the dead on the third day, allowing for the redemption of the whole human race.

Cain, however, was trying to again regain the fellowship with God which his father once enjoyed in the Garden of Eden, as a tiller of the soil. Cain went out from the Garden and began to learn the most efficacious and successful ways to till the ground to bring an increase, and he brought the fruit of the ground to the LORD as his sacrifice.

Later, under the Levitical covenant, God established the system of grain sacrifices and grain offerings, for certain circumstances.

Why was Cain's sacrifice and offering of grain not acceptable?

It had nothing to do with the sacrifice, rather, it had to do with Cain's heart. Cain's reason for bringing the offering of the fruit of the ground before the LORD was to regain entry into the Garden of Eden, the place his father held with God, as a tiller of the ground. Cain was trying to do it his own way, and this was not in line with God's way. God's way from the moment Adam and Eve sinned was through blood atonement, and redemption. Cain was trying to earn his way back by showing the LORD what a good tiller of the ground he had become.

Also, the text claims Cain merely brought of the fruit of the ground while Abel brought of the firstlings of his flock and of the fat thereof. In other words, Cain just brought some grain before the LORD as an offering, while Abel brought the best of his flocks, and one of the fattest of his herd. Abel was offering his best to the LORD whereas Cain was merely offering a sample of his labor.

This is why God did not accept Cain's sacrifice. First because of the attitude of Cain's heart, because no man would ever again fulfill the role of Adam in the Garden of Eden until the Son of God returned to reestablish his kingdom in the Millennial Earth, releasing the Earth from its groaning **Romans 8:22;** for a one thousand year period at the end of the world during which Jesus Himself would reign from Jerusalem over all the Earth.

Cain, trying to be reinstated into the leisurely life of the Garden of Eden, was not accepted, and his countenance fell, which means he became angry. He became the first murderer, and as God prophesied concerning him, sin lay at his door. **Genesis 4:7;** It says Cain could rule over the sin, he was still in the temptation stage at this time, still able to choose not to sin. He was just beginning to feel the insistent influence from Satan in this act of murder, because Satan was a murderer from the beginning. **John 8:44;**

This murder by Cain was an act which came from inside him, from his own heart, his own mind. He could have ruled over it and not done the deed, but he succumbed to its desire and followed through with the sin. And the result was his brother lay dead, and the blood of his brother cried unto God from the ground. **Genesis 4:10;** God hears the cry of blood, when it is spilled, as he heard the cry of the blood of Jesus, when He was crucified on the hill of Golgotha.

Genesis 4:8-12; "And Cain talked with Abel his brother: and it came to pass, when they were in the field," Notice they were both in the open field, Cain tending to the ground, while Abel was tending his flock. ***"...that Cain rose up against Abel his brother, and slew him."*** Some scholars teach when Cain slew Abel he killed one quarter of the Earth's population, but as previously shown, this was not so.

"And the LORD said unto Cain, Where is Abel thy brother? And he said, I know not: Am I my brother's keeper?" God asked Cain a question, and Cain answered quickly and followed his answer with another question, a question which has become the hallmark of the Cain and Abel story throughout our history, "Am I my brother's keeper?"

Immediately God asked him; "What have you done?" God knew, before He asked, Cain had killed his brother, and the answer Cain offered the LORD was proof of two things; First, it was proof of Cain's guilt, and Second, it was proof Cain didn't understand the nature of the LORD, if he thought he could fool Him.

This was the crux of the matter where Cain was concerned, Cain didn't know God very well. He didn't know God and he didn't know God's plan for redemption. His brother Abel did, and Abel realized it was through a blood sacrifice God would again reinstitute fellowship with man.

"And he said, 'What hast thou done?' The voice of thy brother's blood crieth unto me from the ground." The murder of Abel was Cain's blood sacrifice, and it was again not in harmony with God's eternal plan. Cain made mistakes from the first day with God, and his venal efforts served merely to get him castaway as an exiled vagabond and a pariah in the Earth.

"And now art thou cursed from the earth, which hath opened her mouth to receive thy brother's blood from thy hand; When thou tillest the ground, it shall not henceforth yield unto thee her strength; a fugitive and a vagabond shalt thou be in the earth."

Now, Cain was cursed, not only with a mark, but also with a crucial addendum to the curse pronounced upon his father. The ground was cursed for his father's sake, now he would have to work the ground for a meal, but even this was now denied to Cain. He would not be able to work the ground for a meal anymore. He would be a fugitive, a wanderer upon the Earth, and a vagabond, a beggar for a meal, wherever he went in the Earth.

This is why Cain stated everyone finding him would kill him, because everyone would know Cain killed Abel, and they, the people of the Earth, knew the law was in effect already, a law saying *'…a life for a life…'. **Leviticus 24:21;*** Even though the law was not given until Mt Sinai, it was active in the Earth prior to then, as evidenced many times by the people of the Earth, from Adam to Noah to Abraham to Jacob to Moses.

Genesis 4:13-15; *"And Cain said unto the LORD, My punishment is greater than I can bear. Behold, thou hast driven me out this day from the face of the earth; and from thy face shall I be hid; and I shall be a fugitive and a vagabond in the earth; and it shall come to pass, that every one that findeth me shall slay me. And the LORD said unto him, Therefore whosoever slayeth Cain, vengeance shall be taken on him sevenfold. And the LORD set a mark upon Cain, lest any finding him should kill him."*

God set a mark on Cain, to distinguish him from among the inhabitants of the Earth, so no one would slay him, and they would have compassion on Cain, and feed him in consideration of his inability to grow his own food. To speculate on the nature of this mark is unproductive and an exercise in futility, (like trying to understand the nature of the "stones of fire" ***Ezekiel 28:14-16;*** or what the seven thunders uttered ***Revelation 10:3-4;***) since we are not given enough information to know these inscrutable mysteries for certain. Those who claim it was the coloring of his skin to black are not operating with facts, but rather with prejudicial preconceptions.

Regardless of what God intended for Cain in his wandering, Cain entertained his own ingenious designs about provision for his family, in cities, and by crafts and arts.

Also the very presence of some people who would want to kill him implies a populated Earth when Cain was exiled into at nomadic life. This further reinforces the idea of the sixth day creation of man versus the making of Adam and Eve.

Genesis 4:16-17; "And Cain went out from the presence of the LORD, and dwelt in the land of Nod, on the east of Eden. And Cain knew his wife; and she conceived, and bare Enoch: and he builded a city, and called the name of the city, after the name of his son, Enoch."

Cain went out from the presence of the LORD, he left Eden and went east, away from where Adam and Eve were dwelling. From this it is evident, though Adam and Eve were expelled from the Garden of Eden, they stayed very close to the garden's location, and the LORD dwelt there in the Garden, even after they were expelled. When Cain departed to the east of the Garden of Eden, he went away from where the LORD was maintaining His presence.

It doesn't say Cain found his wife in the land of Nod, it only says he went to dwell in the land of Nod and there he knew his wife, and she bare him a son. However, he most assuredly obtained his wife from one of the many peoples alive on the Earth, inhabitants from the offspring of the men and women created in day six. This verse also says from then on, Cain, and his kin, lived in cities. Since they could not be farmers anymore, they developed all sorts of arts and crafts, and flourished in the trades. ***Genesis 4:20-22;*** They were responsible for laying the foundations of what is today referred to as 'civilization'.

They used these arts and crafts, musical instruments and tools, and all implements of living, to barter for food among the inhabitants of the Earth, since they were no longer capable of growing it for themselves. It was a symbiotic relationship serving both sides equally, giving Cain's descendants the necessary food for survival, and vice versa, giving the rest of the inhabitants the tools necessary to make life easier and more enjoyable.

Cain's descendants became proficient at music, art, the trades, tool making and merchandising. It is interesting to note Satan was also one who was proficient in merchandising, ***Ezekiel 28:16;*** and also in trafficking. ***Ezekiel 28:18;***

Another later addendum to the life of Cain is found further on in the text ***Genesis 4:23-24;*** where Lamech, the son of Methusael and a descendant of Cain, killed a man who wounded him, in self-defense. He boasts to his wives, **"If Cain shall be avenged sevenfold, then I shall be avenged seventy and sevenfold."** Although the text is not clear about who Lamech killed, the reference to the vengeance afforded Cain, and the usurpation of said vengeance to himself, elevenfold, makes a discerning mind think Lamech killed Cain.

In support of this is the next two verses which speak of Eve getting a replacement for Abel, whom Cain slew.

Genesis 4:25-26; "And Adam knew his wife again; and she bare a son, and called his name Seth: For God, said she, hath appointed me another seed instead of Abel, whom Cain slew. And to Seth, to him also there was born a son; and he called his name Enos: then began men to call upon the name of the LORD."

The last part of this verse, *"...to call upon the name of the LORD,,,"* where *"call upon"* is an improper rendering from the *Hebrew qara:* קָרָא *to call out to someone, to accost someone, to cry out in anguish.* The way the verse is constructed it leaves the impression men, in the days of Enos, began to turn their hearts toward the LORD again, but the verse is instead saying men began to accost the name of the LORD, or to curse God.

If Cain were still alive, there would not have needed to be a replacement for Abel, but if Cain were dead, then the verses make sense. Seth was a replacement for the two sons Eve lost. This description of the birth of Seth immediately followed the verses concerning Lamech's boast to his two wives about the seven fold vengeance of Cain's being multiplied to him eleven fold more, meaning Lamech killed Cain.

The thought Seth replaced Abel as a righteous representative of God on the Earth is not supported by the text. It says during Seth's son Enos' day, men began to call themselves by the name of the LORD, or to usurp the authority of God for themselves. This is not righteousness, in fact, it may be worse than Cain's sin, since the second commandment prohibits the taking of the name of the LORD in vain, *Exodus 20:7;* and it is the fifth commandment prohibiting murder. *Exodus 20:13;*

Occam's Razor

This begins to clear up some of the anthropological mystery about the early history of hominids on the Earth, and realigns the Word of God in the scientific context, instead of making science at odds with the Word of God. Science can have millennia or even millions of years in which to place hominids and the settlement of humans into rudimentary cities, or even into early civilizations, and the Bible can have the making of Adam and Eve at approximately *4000 BCE*. The dichotomy between science and the Word of God has disappeared with this simple solution, Occam's Razor.

Occam's Razor, or Ockam's Razor is a problem-solving principle attributed to William of Ockham, *CE 1287–1347,* who was an English Franciscan friar, scholastic philosopher and theologian. The principle can be interpreted as stating *"Among competing hypotheses, the one with the fewest assumptions should be selected."*

More Proofs

This is further reinforced by the following:
Genesis 5:1-2; "This is the book of the generations of Adam."
The Book of Genesis; is the book of the generations of Adam, the man God formed to put into the garden, the farmer.

"In the day..." The sixth day of ***Genesis One;*** *"...that God created..."* Hebrew: bara: בָּרָא *"...man, in the likeness of God made..."* Hebrew: yatsar: יָצַר *"...he him;"* He formed Adam in His likeness, He breathed into him the breath of life and he became a living soul. *"...Male and female created..."* Hebrew: bara: בָּרָא *"...he them;"* He created them male and female in day six, yet the forming of Adam was done many, many days before the

forming of Eve, taking into account all the events transpiring in the garden between the forming of the man, and the surgery to form the woman. In the interval between the forming of Adam and Eve, God planted the Garden, and it grew to the point where it needed to be tended. Then God set the man to tend it and make sure it grew. Then God saw Adam was lonely and He brought all the animals one by one to Adam to be named. Then He saw Adam was still lonely so He made Eve. Surely no one can surmise this was all done in the same day, an evening and a morning.

"...and blessed them, and called their name Adam..." Hebrew: adam: אָדָם *ruddy man "...in the day..."* Here God reasserts man was created, male and female, in a day; the sixth day *"...when they were created."*

The creation of man was an event occurring on the sixth day, male and female, and the forming of Adam happened at a later date, and the forming of Eve at a later date still.

A Cardinal Error

When considering Cardinal Ussher's timetable for the creation at *4004 BCE*, in comparison to the events just reviewed, it is easy to deduce what really occurred at *4004 BCE* was the expulsion of Adam and Eve from the garden, not the creation of the Universe, or even the Earth.

Ancient Timeline

A timeline outlining from Adam's expulsion out of the Garden of Eden until Joshua and the children of Israel took the Promised Land would be helpful here:

It is uncertain how long Adam and Eve lived in the Garden of Eden before their inconvenient expulsion. It is evident they lived there for quite some time. Judging from the extended ages of his immediate descendants, it can surely be determined Adam lived a long time in the Garden before he was forced out by God and kept out by the Cherubim with the flaming sword. *Genesis 3:24;* Then he lived for another 930 years on the cursed Earth, after the expulsion from the Garden of Eden.

This can be deduced from three facts; First, Adam and Eve established a pattern of walking with God in the Garden in the cool of the day. *Genesis 3:8;* Second, Adam spent enough time alone, before God made Eve, to grow lonely. Lonely enough for God to see his need for an help meet, *Genesis 2:20-22;* and it takes a long time indeed for loneliness to manifest itself to such an extent, as in Adam's case. Third, Adam took enough time to name all the animals as God brought them to him. *Genesis 2:19-20;*

Here is a chart delineating the ages of the patriarchs, starting at 0–930 for Adam. *Genesis 5:5;* Adam was 130 when Seth was born, so Seth's date starts at 130. All dates are from the expulsion of Adam from the Garden of Eden, which Cardinal Ussher set at *4004 BCE*. This thesis does not dispute his calculation, only his conclusion.

The Timeline of the Patriarchs

0 - Adam and Eve banished from the Garden of Eden: *Genesis 3:24;*
ca. 50 - Cain kills Abel: *Genesis 4:8;* This is the first recorded death of a man.
130 - Seth born to Adam and Eve in place of Cain and Abel: *Genesis 5:3;*
235 - Enos born to Seth: *Genesis 5:6;*
325 - Cainan born to Enos: *Genesis 5:9;*
395 - Mahalaleel born to Cainan: *Genesis 5:12;*

460 - Jared born to Mahalaleel: *Genesis 5:15;*
632 - Enoch born to Jared: *Genesis 5:18;*
707 - Methuselah born to Enoch: *Genesis 5:21;*
894 - Lamech born to Methuselah: *Genesis 5:25;*
930 - Adam dies: *Genesis 5:5;* This was the first natural recorded death of a man since Abel.
1007 - Enoch taken to Heaven by God: *Genesis 5:23-24;* This is the first recorded instance of a man who never died.
1042 - Seth dies: *Genesis 5:8;*
1076 - Noah born to Lamech: *Genesis 5:28;*
1140 - Enos dies: *Genesis 5:11;*
1235 - Cainan dies: *Genesis 5:14;*
1290 - Mahalaleel dies: *Genesis 5:17;*
1422 - Jared dies: *Genesis 5:20;*
ca. 1556 - The "sons of God" began to marry women and giants were born to them: *Genesis 6:4;* This date is reckoned backward from the date of the Flood of Noah 120 years as determined by God. *Genesis 6:3;*
1576 – Ham and Japheth born to Noah: *Genesis 5:32;*
1578 - Shem born to Noah: *Genesis 5:32; & Genesis 11:10;*
1671 - Lamech dies: *Genesis 5:31;*
1676 - Methuselah dies: *Genesis 5:27;*
1676 – All men alive at this event died, except Noah and those in the ark: *Genesis 7:21;*
1676 – Noah's Flood occurs: *Genesis 7:11;* What many people fail to take into account when they read the early chapters of *The Book of Genesis;* is the timing of Noah's Flood, which takes place seemingly early on in the history of man, was actually 1,676 years after Adam left the Garden of Eden. It appears to be an event directly following the exit from Eden because of its place in the first few chapters of the Bible, but it actually occurred almost two millennia later.

1678 - Arphaxad born to Shem: *Genesis 11:10;*
1713 - Salah born to Arphaxad: *Genesis 11:12;*
1743 - Eber born to Salah: *Genesis 11:14;*
1777 - Peleg born to Eber: *Genesis 11:16;*
1807 - Reu born to Peleg: *Genesis 11:18;*
1839 - Serug born to Reu: *Genesis 11:20;*
1869 - Nahor born to Serug: *Genesis 11:22;*
ca. 1897 - The Earth is divided in the days of Peleg: *Genesis 10:25;*

In the days of Peleg, two hundred and twenty one years after the flood, the continents started their drift (continental drift, or plate tectonics), and were divided. When the water subsided off the Earth, after the flood, the water went into subterranean cavities which caused a lubrication effect on the continental plates dislodging them from their entrenched positions and caused the beginning of the movement of the plates, called plate tectonics.

ca. 1897 – The Tower of Babel: *Genesis 11:7-8;* Man's languages are confounded and men are scattered abroad upon the face of all the Earth. First, men are scattered abroad upon the face of all the Earth, or across the many divided lands, and then their languages change, or are confounded.

1898 - Terah born to Nahor: *Genesis 11:24;*
1968 - Abram, Nahor and Haran born to Terah: *Genesis 11:26;*
2016 - Peleg dies: *Genesis 11:18-19;*
2017 - Nahor, Serug's son dies: *Genesis 11:24-25;*
2026 - Noah dies: *Genesis 9:29;*
2046 - Reu dies: *Genesis 11:20-21;*
2054 - Ishmael born to Abraham: *Genesis 16:16;*
ca. 2067 - Sodom and Gomorrah are destroyed by fire and brimstone from Heaven: *Genesis 19:24;*
2068 - Isaac born to Abraham: *Genesis 21:5;*

2069 - Serug dies: *Genesis 11:22-23;*
2103 - Terah dies: *Genesis 11:32;*
2116 - Arphaxad dies: *Genesis 11:12-13;*
2128 - Esau and Jacob born to Isaac: *Genesis 25:25-26;*
2143 - Abraham dies: *Genesis 25:7;*
2146 - Salah dies: *Genesis 11:14-15;*
ca. 2175- Levi born to Jacob: *Genesis 29:34;*
2178 - Shem dies: *Genesis 11:10-11;*
2191 - Ishmael dies: *Genesis 25:17;*
2207 - Eber dies: *Genesis 11:16-17;*
2221 - Joseph born to Jacob: *Genesis 30:23;*
2248 - Isaac dies: *Genesis 35:28;*
ca. 2258 - Israel and his household go down into Egypt to sojourn until the great famine departed: *Genesis 47:9;* It was during the next seven years of hardship the Earth, or at least the Levant, suffered the tremendous famine of the days of Joseph as he predicted in his interpretation of Pharaoh's dream.
2275 - Israel dies in Egypt: *Genesis 47:28;*
ca. 2310 - Kohath born to Levi: *Genesis 46:11;*
ca. 2312 - Levi dies: *Exodus 6:16;*
2331 - Joseph dies: *Genesis 50:26;*
ca. 2340 - Amram born to Kohath: *Exodus 6:18;*
ca. 2473 – Aaron (Moses brother and First High Priest) born to Amram: *Exodus 6:20; & Exodus7:7;*
ca. 2476 to 2608 - Moses born to Amram: *Exodus 6:20;* Two dates are given on four of the next six entries because the calculation from the expulsion from the garden brings the first number and the time mentioned as the length of time the Israelites spent in captivity (430 years) brings the second number. This is not precise.
ca. 2443 - Kohath dies: *Exodus 6:18;*
ca. 2477 - Amram dies: *Exodus 6:20;*

ca. 2556 to 2688 - Moses leads the children of Israel out of Egypt: ***Exodus 12:40-41;***

ca. 2596 to 2728 - Moses dies: ***Deuteronomy 34:5-7;***

ca 2650 to 2782 - Joshua takes the Promised Land: ***Joshua 1:1; to Joshua 24:33;***

Joshua, and the Israelites, spent fourteen years conquering the Promised Land. Forty years earlier, when the spies went up to spy out the land, the report they returned with described the Promised Land as a *"...land flowing with milk and honey." Exodus 13:5;* The reason they didn't go immediately into the land was because there were giants in the land spotted during the recognizance visit to the land by the twelve spies. ***Numbers 13:32-33;*** What happened to the giants when Joshua conquered the land just forty years later? What happened to the *"land flowing with milk and honey",* the state of the land just forty years earlier?

The Sons of God

A long time ago, in the days before Noah's Flood, and shortly after it, some angels came down to Earth and cohabited with some human women. The Bible refers to the angels as the "sons of God" in these verses, **Genesis 6:1-4;** and leaves no doubt this was not a normal union, approved by God.

These angels married human women and they bore children with them. Their offspring were the giants, *Hebrew nephilim:* נְפִלִים *men of renown; giants;* and these giants propagated and continued to propagate until the last of them were destroyed in the days of the kings of Israel.

The Bible tells us a great deal about the "sons of God".

The Old Testament describes them as angels and to correctly understand the nature of these beings from the descriptions in the Old Testament, a better understanding of the whole plan of God, from before the beginning, to after the end, may be obtained.

The New Testament also has many descriptions of the "sons of God". The emphasis in the New Testament is not on angels, it is on the saints, and this difference was deliberate by God, to show the nature of the angels and how their nature relates to believers in this Dispensation, and, moreover, into the next. It is very possible the angels were the saints in the last dispensation, as the saints will be the angels in the next one.

The Old Testament

Some scholars say the "sons of God" of **Genesis 6;** are demons which roam the Earth doing the will of their master, Satan. However, the demons are the remains of

another group of beings, and the subject of a separate study.
See: The Great Tribulation and the War in Heaven;

Other scholars say the "sons of God" are men of the godly line of Seth, who come in unto the daughters of men whom, they say, come from the ungodly line of Cain. In light of all the opposing evidence pointing to the "sons of God" as angels, and the lack of evidence to support this claim, it is inconceivable they continue this claim.

This study will show the Bible clearly identifies the New Testament references to the "sons of God" as angels.

These "sons of God", were; *Jude 6; "...the angels which kept not their first estate, but left their own habitation..."* and are now *Jude 6; "...reserved in everlasting chains under darkness unto the judgement of that great day".*

They are the *II Peter 2:4; "...angels that sinned...".* And God *"...cast them down to hell..."* Greek: tartaros: *Τάρταρος* the abyss, hell *"...and delivered them into chains of darkness, to be reserved unto judgement;"* Because of their sin, these angels, who took wives of women and birthed offspring, the giants of old, are locked away in Tartarus until the judgement day. It seems obvious from this verse, these beings and their deeds, were not approved of God. Jude clearly identifies them as angels, not demons or men from the godly line of Seth.

They were, in fact, part of a whole class of angels who were not fully under the will of God, some of whom later fell with Satan, *Revelation 12:4;* at the time of the War in Heaven during the incarnation of Jesus on the Earth, and who are now the principalities and powers of the air mentioned in scripture. *Ephesians 1:21;*

They were placed in positions of authority over the nations of the Earth and many times they have an agenda all their own, an agenda which is in opposition to the will of God. ***Daniel 10:13; "But the Prince of the Kingdom of Persia withstood me one and twenty days: But, lo, Michael, one of the chief princes, came to help me; and I remained there with the kings of Persia."***

The Prince of the Kingdom of Persia mentioned here, is obviously not the Earthly ruler of Persia, because the Earthly ruler of Persia would not have hindered the messenger of God in his appointed task. This Prince of Persia is obviously an angel, one of the principalities and powers of the air over the Earthly Kingdom of Persia. ***Ephesians 1:21;*** Moreover, this Prince has proven himself to be operating out of line with the precepts of the LORD God, because he withstood the messenger of the LORD from delivering the message to Daniel, a task appointed to him from the day Daniel set his mind to understand. ***Daniel 10:12;***

If this Prince of Persia withstood the messenger, then he was an angel whose agenda was not in concert with God's and he falls into the same category as the angels in Noah's day, who were called in scripture the "sons of God". This is another group of angels whose agenda was not in concert with the LORD's, as evidenced by the severity of the punishment God meted out to them for their actions.

What does it mean when it says they ***Jude 6; "...kept not their first estate, but left their own habitation..."*** and why was this bad enough to require them to be ***"...locked away in everlasting chains under darkness until the judgement of that great day"***?

To understand this completely it is necessary to look at everything the Old Testament has to say about the "sons of God".

Genesis 6:1-2; "And it came to pass, when men began to multiply on the face of the earth..."

Many people do not realize there was 1,676 years from the time when Adam was expelled from the Garden of Eden to the days of the Flood. Men, during this span of time, multiplied upon the face of the Earth, and all during this time, these angels, the "sons of God", took human women as wives.

"...and daughters were born unto them..."

This declares there were daughters being born to men all during this time when men began to multiply upon the Earth. This not only includes the offspring of Adam and Eve, it also includes the offspring of the men created male and female in day six of the Reconstruction. Adam and Eve also assuredly sired daughters, although the Bible does not specifically indicate this in their case. It is probably from these descendants of the day six man where Cain found his wife before they moved to the Land of Nod, east of Eden.

Men lived an exceedingly long time in those days, Adam lived nine hundred and thirty years after he was expelled from the Garden of Eden, and it is uncertain how long he lived while he was tending the Garden. Methuselah lived 969 years, the oldest on record, and even in the days of Noah, who himself lived to nine hundred and fifty years, men were still living long lives. From Adam's day to Noah's day, a period of 1,676 years, the life expectancy stayed stable, but in Shen's day it decreased by over three hundred years and Shen was only six hundred years old when he died, and Arphaxad, Shem's son lived just four hundred and eighty five years. Shortly after Noah's day, the life expectancy began to rapidly decrease and by the days of Abraham, just 400-450 years later, it was down to a mere 175 years. These long ages, as previously shown, were a result of the bubble of water from day two of the Reconstruction enclosing the Earth in the upper atmosphere.

Needless to say, the offspring of these men of the pre-diluvia age were most likely numerous indeed. Considering the gestation period of a human woman is nine months and the modern embryo carrying capacity of a human womb is between twenty and thirty fetuses, and with the length of lives in those days, the carrying capacity was probably tremendously more, maybe as many as one hundred to two hundred per couple. With this in mind, there was prospectively a great population of these *"giants in the earth in those days."*

"...that the sons of God saw the daughters of men that they were fair; and they took them wives of all which they chose."

This was the abomination causing them (the angels identified as the "sons of God") to be locked away in chains of darkness until the judgement day. *Jude 6;* This is what is meant when it says they *"...kept not their first estate..."* and *"...left their own habitation...".*

The first estate of an angel is the spiritual realm. As is evidenced from numerous scriptures throughout the Bible, angels have the ability to also operate in the physical realm. The physical realm is not their first estate though, and when some of the angels came down here and copulated with women, they left their first estate and became physical bodies. In order to do this, they necessarily left their own habitation, which is Heaven, and the spiritual realm, and came and dwelt in the physical realm.

God has reserved these angels in chains until the final judgment day for these offenses,

Jesus said the angels which are in Heaven do not marry, *Matthew 22:30;* and *Mark 12:25;* but these angels violated this stricture and left their first estate, left their habitation and came here and cohabitated with human women.

Genesis 6:3; "And the LORD said, My spirit shall not always strive with man, for that he also is flesh…"

This says much about the tri-fold nature of man. If man was only a spiritual being, as is God, ***John 4:24;*** then he would never die, as long as he possessed God's spiritual nature enlivening his own. God is stating, here, the man is also flesh, and what is more, since the sin nature was introduced, man is ruled by the flesh. And the desire of the women back then to marry angels and have their children is a statement saying the flesh was winning the battle in man and the spirit side of man would need redemption. It was an attempt by man, in his fleshly desires, to regain the spiritually dominant side of his nature by close intercourse with predominantly spiritual beings. It was an abomination to God, and God dealt with it in short order.

"…yet his days shall be an hundred and twenty years."

God locked the offending angels away into Tartarus, another compartment of the underworld along with Hades (Hell) and Paradise, and limited the number of years a man would live.

This was a twofold punishment on both the angels who sinned and the men, or women, who sinned. This explains the limitation of the length of a man's life in spiritual terms, seeing the life spans of man after Noah's Flood quickly diminished to, and rarely exceeded, 120 years. An explanation of the physical reason man's life span decreased rapidly after the flood is found in the section on the Agent of Mass Destruction, but this was the other side of the coin, the spiritual reason for the punishment.

There is another meaning to this part of the verse, because God would allow man another one hundred and twenty years, and then He would send Noah's Flood. From the time this pronouncement was made, there was another one hundred and twenty years before the flood hit the Earth, in the days of Noah.

Genesis 6:4; "There were giants..." *Hebrew nephilim:* נְפִלִים *men of renown; giants; "...in the earth..."* These giants were the offspring of the angels and human women. This bespeaks the awesome power of the spirit essence exemplified in a physical union of the two species, how deliberately separate angels and man are as different species. Unions of this nature produce men of giant stature and power, whereas a union of man to man produces only other men.

Some scientists doubt the existence of giants in the Earth because, they say, if they existed where are their bones? Apparently these scientists are not fully informed, for there are giant bones being discovered all over the world, even here in America. In a cabinet in a back room of the Hunboldt Museum in Winnemucca, Nevada are kept the bones in question. They are not out prominently on display for reasons which have to do with security and a desire on the part of the curator not to offend local Indians. Few are allowed to see them, however, they are there and informed scientists should be aware of this.

"...in those days; and also after that..." What days? Well, the days in question here are the days of Noah, the days of the Flood, and the days preceding and following Noah's Flood. This verse is saying *"...in those days..."*, the days of Noah before the flood, *"...and also after that...",* the days after the flood, this union of angels and women happened. It happened once before the flood, and God destroyed all those beings, only saving Noah and his

family, and then when the flood ended and men began to multiply again on the Earth, those angels did it again, and more giants were born to the women. However, this time, God dealt with the angels and locked them up until the judgement day. ***Jude 6;*** Most of the giants born in the days before the flood were destroyed by the flood, and the ones born after the flood are the ones who continued to propagate until the last ones were defeated in the days of the kings of Israel. There were of necessity far fewer born after the flood than before it, based on the narrative of the account, and this was why God dealt with the angels by locking them away after the flood. The punishment of the flood was not enough to discourage this act by both the angels and by women, and when they continued the practice after the flood, God saw the necessity of more severe punishment for the offense, both locking up the angels, and reducing man's attained age to one hundred and twenty years.

And as for the giants born of this second union, some of their offspring, or perhaps some of the very giants who were born of angels themselves, were still around when the spies under Moses went up to spy out the land at the Exodus, over one thousand years later. ***Numbers 13:33;***

Goliath was also one of their offspring in the days of David.

See our section on: The Giants in the Earth.

"...when the sons of God came in unto the daughters of men, and they bare children to them, the same became mighty men which were of old, men of renown." Hebrew: enowsh: אֱנוֹשׁ *men of renown, mortal.*

These giants, mighty men, men of renown, were mortal. Being the offspring of angels, spiritual beings, and women, physical beings, caused them to be mortal whereas their fathers were immortal beings, the angels.

This calls to mind the heroes of Greek mythology, of Achilles, of Hercules, of Perseus, of the Titans, and others, who were the children of the gods and humans, possessing the hearts of the gods, but were mortal. Perhaps this mythology's roots are anchored in the giants. **Genesis Six;**

These giants also give us a possible answer to another enigma, those whose names were not written in the Book of Life from the foundation of the world. In **Revelation 17:8;** and **Revelation 13:8**; it informs us there are those whose names are not written in the Book of Life from the foundation of the world. It is known from **Revelation 20:15;** anyone whose name is not found written in the Book of Life at the Great White Throne Judgment is thrown into the lake of fire. It is further known men are exempt from the lake of fire because of the redemptive work of Jesus,

Who are the ones whose names are not written in the Book of Life? Perhaps these giants, who were part man, part angel, are the ones.

Genesis 6:5; "And GOD saw that the wickedness of man was great in the earth, and that every imagination of the thoughts of his heart was only evil continually."

This is a sad description of the state of man in those days. Every imagination of the thoughts of his heart was only evil continually. In other words, men thought evil always, and never stopped thinking evil, moment to moment, day after day. Always. As bad as it is in the world today, and it keeps getting progressively worse as each day passes, men are not even close to the state they were in during the days before the flood. Even the most hardened criminal has moments when his thoughts are not evil, and surely the average person has many more of those moments when goodness and purity occupy his thoughts, than those when evil intentions creep in unawares. In the days before Noah's Flood, the Bible says the wickedness of man was great in the Earth. All over the Earth, men were wallowing in evil, and EVERY imagination of his thoughts were ONLY evil CONTINUALLY. This is an inconceivable state of affairs. It is unimaginable how man could have gotten to such a sorry decrepit and derelict state of affairs in such a short period of time. Remember, those people did not have the convicting power of either the written law, as did those from Moses to Jesus, or of the Holy Spirit, as do those from Jesus to today and into the future. Man's conscience alone is insufficient to overcome evil tendencies. He needs God.

Genesis 6:6; "And it repented the LORD that he had made man on the earth, and it grieved him at his heart."

No wonder God was repentant for making man, as one might be repentant of a sin, and He was grieved at His heart, because of man. Why then did He make man?

Man was the great battle ground, the grand experiment. God was looking for fellowship. He was looking for a creature to have fellowship with, and He decided upon man.

For dateless ages past God enjoyed fellowship with the angels, from His first Dispensation, but they proved defective, and evil was too powerful a force for even some of them to overcome by themselves. The crown of His creation, Lucifer himself, was overcome by evil, and it consumed him, *Ezekiel 28:16-18;* until the latter end of the creature was unrecognizable to God. And many more of the angels He created also rebelled and took the path of evil instead of staying true to God, to goodness.

This time, God would make a perfect (mature) creation, and this was man. God realized this time the force of evil was too strong for any creation of His to overcome within themselves. This time around He made a way of redemption for His creation, and this time He could have an everlasting eternity with them, fellowshipping with them, and He Himself would have to be the one to provide redemption. He did this by the sacrifice of His Son, Jesus.

Genesis 6:7; "And the LORD said, I will destroy man whom I have created from the face of the earth; both man, and beast, and the creeping thing, and the fowls of the air; for it repenteth me that I have made them."

At this juncture in man's history, God purposed to destroy man from the Earth, and the beasts also, and every creeping thing, and the birds of the air. The corruption of the creation was complete, even to the animals, and the fowl of the air. He couldn't have eternal fellowship with them if He destroyed them completely.

Genesis 6:8; "But Noah found grace in the eyes of the LORD..."

Noah was saved, and his family, eight souls, and he was told to save the animals, two by two of unclean beasts, *Genesis 6:19-20;* and seven by seven of clean beasts. *Genesis 7:2-3;*

The Job Narrative

To continue the study on these beings, these "sons of God", it is necessary to look to other parts of the Bible, particularly *The Book of Job;* In *The Book of Job;* it speaks of these "sons of God" in three places:

First in *Job 1:6; "Now there was a day when the sons of God came to present themselves before the LORD, and Satan came also among them."*

All the angels are referred to as "sons of God" in these verses in the Old Testament. And one of the requirements of these angels is they present themselves before the LORD. How can anyone say references to the "sons of God" are references to demons, with this verse in mind, did indeed the demons have a requirement to present themselves before the LORD? No, but angels did. And did the men from the godly line of Seth have a requirement to present themselves before the LORD? No, but angels did.

Second in ***Job 2:1; "Again there was a day when the sons of God came to present themselves before the LORD, and Satan came also among them to present himself before the LORD."***

This verse, which seems redundant in light of the verse above, is not, however, redundant. This verse again tells us the angels were required to present themselves before the LORD. It gives us further insight as to these requirements, because this was a different day than the one in ***Job 1:6;*** and therefore shows us the angels are required to present themselves before the LORD regularly, maybe even on a schedule.

The fact Satan is mentioned separately from the "sons of God" indicates Satan holds a different position from the angels, which indeed he does. Satan, being a fallen angel, was singled out by God in these exchanges, and God asks him where he has been.

This also proves, looking at the chronology of events in the right order, the third of the angels which Satan caused to fall, ***Revelation 12:4;*** were not yet fallen at this time, at the time of the patriarch Job. If they were fallen at this point, God would have asked Satan where he and his angels were, however, God only asked Satan where he was, asking nothing about his angels.

And third in ***Job 38:4-7; "Where wast thou when I laid the foundations of the earth? declare, if thou hast understanding. Who hath laid the measures thereof, if*** *thou knowest? or who hath stretched the line upon it? Whereupon are the foundations thereof fastened? or who laid the corner stone thereof; When the morning stars sang together, and all the sons of God shouted for joy?"*

These scriptures tell us much about the nature of the creation of the Earth, in the beginning, and of the nature of the realm of the spirit beings (angels). It tells us there were foundations laid in the creation of the Earth, very much like a builder would lay a foundation when erecting a building. In fact, the terminology of this creation event is identical to the terminology used in the construction of a building. Who laid the measures thereon? Who stretched the line upon it? Where are the foundations fastened? Who laid the cornerstone? The construction of a planet, on a celestial scale, is very akin to the construction of a building, on a terrestrial scale.

What do these verses tell us about the realm of spirit beings?

When the Earth was created, in the beginning, the morning stars sang together, and all the "sons of God" shouted for joy. The angels and the morning stars (plural) existed before the beginning. ***Genesis 1:1;*** It was in the beginning the foundations of the Earth were laid.

Who are these morning stars?

This reference is to a plurality of morning stars, how many were there?

Is there any significance to the fact they were singing?

There are two beings referred to, in scripture, as morning stars; Jesus; ***Job 38:7; Revelation 2:28; and Revelation 22:16;*** and Satan; ***Job 38:7; and Isaiah 14:12;***

Jesus was not the only spiritual being referred to as one of the morning stars, Satan was also referred to as such before the beginning. In fact, his place was equated with

Jesus, since they both were singing together, and the sense of the verse is they were doing this singing as part of a ceremony, a celebration, as was the shout for joy of the "sons of God", a celebration of the creation by the Creator.

Now to analyze these "sons of God" and what this is telling us about them.

"...all the sons of God shouted for joy."

Since, as established, the "sons of God" were angels, this verse tells us all the angels were shouting for joy at the creation of the Earth. They were all in harmony, in unity, and in concert with their LORD.

When it says all were in harmony, it includes, the angels which kept not their first estate, but sinned, an event which occurred much later at the time of Noah's Flood, *Genesis 6:4;* and the angels who were the principalities and powers of the air, *Ephesians 1:21;* of the nations who later carried on their own agenda *Daniel 10:12;* and the third of the stars of Heaven (angels) who fell with Satan at the War in Heaven. *Revelation 12:4;* It also includes the angels who stayed loyal to God and are now sent to us as ministering spirits. *Hebrews 1:14;* It also includes Satan, who as Lucifer, one of the morning stars, who was singing in concert with the other morning star, Jesus.

He was not yet fallen at this point.

The New Testament

Now it is needful to examine the "sons of God" as described in the New Testament, and see the transformation of the description from angels to the saints of the Most High. This transformation is not by accident. The LORD is allowing all to see, through this description, the place of the saints, ("sons of God"), throughout this Dispensation, and into the next, with a proper understanding of the nature of the "sons of God" from the last Dispensation, into this one, from before the beginning, angels.

Becoming a Son of God

John 1:12; "But as many as received him, to them gave he power to become the sons of God, even to them that believe on his name:" To gain the power to become one of the "sons of God", in the New Covenant, it is necessary to receive Jesus as the savior and LORD of your life and believe on His name and believe in your heart God raised Him from the dead. *Romans 10:9;* This is the allotment of the saints, those who have given their life to Jesus, and who follow Him by taking up His cross daily. *Luke 9:23;*

Romans 8:14; "For as many as are led by the spirit of God, they are the sons of God." This is another way to be classified as one of the "son of God", by being led by His Spirit, one becomes one of the "sons of God".

Romans 8:19; "For the earnest expectation of the creature waiteth for the manifestation of the sons of God." The saints are not yet manifested as the "sons of God". They will be at His Second Coming, when the earnest expectation of the creature will be realized, and they will learn the skills necessary to take their place as the ministering spirits unto the next Dispensation.

Philippians 2:15; "That ye may be blameless and harmless, the sons of God, without rebuke, in the midst of a crooked and perverse nation, among whom ye shine as lights in the world," If the saints are to become the "sons of God", they should already be showing signs of this manifestation, being blameless, harmless, without rebuke, and shining as lights in a crooked and perverse world. And be sure, before they are allowed to be the ministering spirits to the next Dispensation, they must put on these blameless and harmless characteristics, and become the true "sons of God" in all righteousness.

1 John 3:1; "Behold, what manner of love the Father hath bestowed upon us, that we should be called the sons of God: Therefore the world knoweth us not, because it knew him not." The saints are the "sons of God" because of the love the Father has bestowed upon them, and this love is in the form of Jesus Christ, whom the world knew not, and therefore the world will not know them either. Believers will be heirs, and joint-heirs, with Christ, heirs of the kingdom. **Romans 8:17; Hebrews 1:2; James 2:5;**

1 John 3:2; "Beloved, now are we the sons of God, and it doth not yet appear what we shall be: But we know that, when he shall appear, we shall be like him; for we shall see him as he is." This says we are the "sons of God", but it doesn't yet appear what we shall be, but when He comes, we shall see Him as He is, and be like Him, and then it shall appear what we shall be, the heirs with Him in the kingdom, and the anointed ministers to those who would come in the next Dispensation.

John Sees a Prophet

The following verses from ***The Book of Revelation;*** speaks about one of the angels, who was showing John the holy city, New Jerusalem, and John fell to worship at the feet of this magnificent angel, when the angel tells him, do not worship me, for I am one of you, one of the prophets.

Revelation 19:10; "And I fell at his feet to worship him. And he said unto me, See thou do it not: I am thy fellow-servant, and of thy brethren that have the testimony of Jesus: worship God: for the testimony of

Jesus is the spirit of prophecy." This 'angel' has the testimony of Jesus, he is a Christian, one of John's brethren. He is a prophet, and informs John the testimony of Jesus is the spirit of prophesy, letting John know he is one of the prophets, maybe Elijah. John was conversing with this 'angel' for some time, and the angel was showing him the things to come, when John felt the need to worship this angel. As he was falling on his knees to worship him, he received this rebuke, and the true nature of the 'angel' was revealed.

And again: ***Revelation 22:8-9; "And I John saw these things, and heard them. And when I had heard and seen, I fell down to worship before the feet of the angel which shewed me these things. Then saith he unto me, see thou do it not: For I am thy fellowservant, and of thy brethren the prophets, and of them which keep the sayings of this book: Worship God."*** This was maybe the same 'angel' as in ***Revelation 19:10***;. Here is another instance when John bent to worship an 'angel' and was rebuked. The 'angel' tells John he was one of his fellow-servants, a man just like John. Yet John mistook him for an angel, and John was not mistaken at all, for even though he asserts himself a man like John, it appears he was also an angel.

Allow me to explain:

A Bold Hypothesis

As already shown, the angels (who are called the "sons of God" in the Old Testament) existed before the beginning. ***Job 38:7;***

Also, the saints of the Most High, (who are called the "sons of God" in the New Testament), are equated to

being an angel in the New Heaven and the New Earth.
Revelation 22:8-9;

The beings who are now known as angels in this Dispensation, were God's creation before this Universe was created. They were the beings who fellowshipped with God in the previous Dispensation, in a previous Universe. Every Big Bang/Big Crunch, creates a different Universe, and a different Dispensation. It is possible there could be an everlasting parade of Universes/Dispensations in which God is filling the creation, the place where he dwells, Heaven.

The angels today could very well be, based on this hypothesis, the inhabitants, maybe even the redeemed, from the last Dispensation/Universe, and it is impossible to tell if there were any Dispensations before theirs. These angels took on the role of ministering spirits to the redeemed in this Dispensation.

Whether the angels, in the last Dispensation were similar to us in this Dispensation, requiring redemption before they moved into this Dispensation as ministering spirits to God's people, the redeemed of this Dispensation, is not clear. However, from the available evidence, there appears to be no redemption for the fallen angels of the last Dispensation, Their place is reserved in the Lake of Fire.
Matthew 25:41;

It is possible their Dispensation came to an end, as will this one. In this one, Satan, a highly exalted being from the last Dispensation, fell. Creation, redemption and sanctification seems to be the processes God uses to cleanse His creation from the stain of sin, which had so pervasively been introduced into His works at Satan's fall. This makes our Dispensation a special case, with a specific purpose, to cleanse it from the stain and stench of sin.

It is possible this process of creation is an endless process from one glorious Dispensation to the next, and as each progressive Dispensation comes and goes, it adds to God's pleasure, and to the host who in the end have fellowship with Him.

If this last scenario is true, then the redeemed of this Dispensation will be the beings, along with the current assemblage of good angels, who have fellowship with God in the next Dispensation. The saints of this Dispensation will be the ministering spirits (angels) in the next Dispensation, in the next Universe. The beings in this Dispensation are the next beings to enter the spiritual stage, in a possible never ending pageant of beings which it pleases God to create, beings who will be the next object of God's endless love.

God gives a minute glimpse of this in the following verses:

Isaiah 42:8-9; "I am the LORD: that is my name: and my glory will I not give to another, neither my praise to graven images. Behold, the former things are come to pass, and new things do I declare: before they spring forth I tell you of them."

God here tells us *"...the former things are come to pass...",* or in other words, our Dispensation has come to pass, at the New Heaven and the New Earth, *"...and new things do I declare...",* or God will again make new things, or new beings in a new Dispensation. And *"...before they spring forth I tell you of them."* God is describing them before He brings them forth.

Isaiah 66:22-23; "For as the new heavens and the new earth, which I will make, shall remain before me, saith the LORD, so shall your seed and your name remain. And it shall come to pass, that from one new moon to another, and from one sabbath to another, shall all flesh come to worship before me, saith the LORD."

Here, God says after the New Heaven and the New Earth are established, there will come a time when all flesh will come to worship Him, from one new moon to another, and one galaxy to another. The implication here is not a new moon in the sense of a new moon, month to month, since the Earth, and also the Moon, were already melted with fervent heat. This new moon is referring to another moon, and also referring to an endless series of moons, and an endless series of Sabbaths, thereby implying an endless series of Dispensations, and an endless series of abiding by the laws (Sabbaths) which implies redemption being necessary again. It is in these endless Dispensations where God can create an endless series of beings (flesh), which will come and worship before Him.

Jesus also referred to these other beings from these endless other Dispensations in:

John 10:16; "And other sheep I have, which are not of this fold: them also I must bring, and they shall hear my voice; and there shall be one fold, and one shepherd."

These other sheep are not of the fold of which men are a part, just as men are not a part of the fold of the angels, the inhabitants of the last Dispensation. The reference here is to sheep of the fold of the next Dispensation, and they also shall hear His (Jesus') voice, and He shall also bring them.

But bring them where? To come and worship before the Father. *Isaiah 66:23;*

Hebrews 2:5; "For unto the angels hath he not put into subjection the world to come, whereof we speak."

Here the scriptures say the world to come, indicating there would be another world to follow this one, would be put unto subjection to Jesus. And the saints will be joint heirs with Him in the management of the world, just as the angels are in management with Him over this one. He informs us the *"...world to come..."* is not put into subjection to the angels, this world was put into subjection to them. The next world will be put into subjection to the saints, the heirs of salvation.

If God's people are the equivalent of the angels in the next Dispensation, does this mean there will be a falling away from among them, a Satan among them? This thesis has shown how all men from this Dispensation gain entry into the next Dispensation, some lame and halt, **Matthew 5:30;** others intact, and since some from this Dispensation are unrepentant, and outside the city reside all sorts of sinners**, Revelation 22:15;** it is a logical assumption to conclude there will again be a disruption of evil in the next Dispensation, although it will probably have a different history than this one. Remember, Satan was God's right hand man in the last Dispensation, he was not one of the wayward sinners, until iniquity was found in him, in this Dispensation/Universe. **See: The Apotheosis of the Apocalypse.**

The being who plays the part of Satan in the next Dispensation may indeed come from God's right hand again. It won't be Satan, however, since he will be spending eternity in the Lake of Fire. And if there is a falling away in the next Dispensation among the saints from this dispensation, it is logical no redemption will be available to those who fall away then, just as there is no redemption for Satan and his angels from the last dispensation who fell away in this one.

The author of this work hopes, beyond hope, this is not the case. One Satan is enough for all Dispensations and Universes.

See: The Apotheosis of the Apocalypse.

The Giants in the Earth

Now since an introduction to the giants has been accomplished, a further analysis of these beings is in order.

Genesis 6:1-2; "And it came to pass, when men began to multiply upon the face of the earth, and daughters were born unto them, That the "sons of God" saw the daughters of men that they were fair; and they took them wives of all which they chose."

Genesis 6:4; "There were giants in the earth in those days; and also after that, when the "sons of God" came in unto the daughters of men, and they bare children to them, the same became mighty men which were of old, men of renown."

All the giants mentioned in scripture have their origin from this starting point. The angels of God came in unto the daughters of men and there were giants born of those unions. This occurred both before the Flood of Noah's days, and after the days of Noah's Flood, when the flood ended and men began to multiply upon the face of the Earth again. This explains why there were giants again in the Earth after the flood, during the Exodus and even until the days of King David.

The Earth before the flood may have been largely populated by these giants since there was 1,676 years from the expulsion of Adam from the Garden of Eden to the Flood in the days of Noah. The description of these giants in scripture indicates these unions between "sons of God" and women was occurring from the Expulsion to the Flood, *"...when men began to multiply upon the face of the earth..."* Men began to multiply when Adam and Eve birthed Cain, and the propagation of the human race during this 1,676 years could have produced numerous gigantic descendants indeed.

This union of women with the "sons of God" was probably the source of the *"...only evil continually..."* in men's hearts, **Genesis 6:5;** at the time of Noah's Flood, the very thing causing God to desire the end of mankind before Him, the cause of His repentance at making man.

The Valley of the Giants

This is a valley on the south side of the seven mountains upon which Jerusalem sits, from the vale of Hinnom in the north to the side of Jebusi in the south and all the way to Enrogel.

The giants born before Noah's Flood, of which we have no record of any individuals, just as a group, they all died in the flood, and may have been the main reason for the flood.

In the days after the Flood, when these "sons of God" again took wives of women, replenishing the giant's numbers on the Earth at the same time men were propagating, the descendants (the giants) took up residence in this valley of the giants, close to the area which would become the seat of the chosen people of God, the nation of Israel.

Joshua 15:8; "And the border went up by the valley of the son of Hinnom unto the south side of the Jebusite; the same is Jerusalem: and the border went up to the top of the mountain that lieth before the valley of Hinnom westward, which is at the end of the valley of the giants northward:"

Joshua 17:15; "And Joshua answered them, If thou be a great people, then get thee up to the wood country, and cut down for thyself there in the land of the Perizzites and of the giants, if mount Ephraim be too narrow for thee."

> *Joshua 18:16; "And the border came down to the end of the mountain that lieth before the valley of the son of Hinnom, and which is in the valley of the giants on the north, and descended to the valley of Hinnom, to the side of Jebusi on the south, and descended to Enrogel,"*

Many of the giants which were the result of the unions of women and the "sons of God" after the Flood, were destroyed by King Chedorlaomer, *Genesis 14:5-7;* who in turn was later killed by Abram and his associates in his quest for the return of Lot, whom the four armies had kidnapped. *Genesis 14:17;*

Of the giants King Chedorlaomer left alive, most of these were annihilated by Joshua and the invading Israelites during the conquest of Canaan.

However, Joshua didn't kill them all and these giants, in later generations, were destroyed by King David, and his mighty men and the rest of the Israelites, during the establishment of Jerusalem as the capital of God's chosen people.

The Families of the Giants

These were descendants of the giants from the unions of the "sons of God" with the daughters of men after the Flood. The following is a detailed list of these giants and their families.

There are many giants mentioned in scripture by name, or by physical characteristic, or by family name, a name which is derived from the name of the original giant, or perhaps from the original "sons of God" whose offspring they were, according to their name.

The Rephaim

The Rephaim were one of these family groups, descendants of the giant Repha. They were inhabitants of the valley of the giants. They were mostly destroyed by Chedorlaomer, King of Elam, and the three Kings with him, just before he defeated the Kings of Sodom (Bera) Gomorrah (Birsha), Admah (Shinab), Zeboiim (Shemeber) and Bela (Zoar) and took Lot captive. Abram slaughtered King Chedorlaomer *Genesis 14:17;* and the three Kings (Arioch, Amraphel and Tidal) for this affront.

The Rephaim resided in the valley of the giants southwest of what would eventually become Jerusalem.

Genesis 14:5; "And in the fourteenth year came Chedorlaomer, and the kings that were with him, and smote the Rephaims in Ashteroth Karnaim, and the Zuzims in Ham, and the Emims in Shaveh Kiriathaim,"

II Samuel 23:13; "...And the troop of the Philistines emcamped in the Valley of the Rephaim."

The Zuzim

The Zuzim were descended from Zuzi. They dwelt mostly in Ham, which is believed to be in North Africa. These were also destroyed by Chedorlaomer, King of Elam, in the battle of the Five Kings against the Four Kings.

Genesis 14:5; "And in the fourteenth year came Chedorlaomer, and the kings that were with him, and smote the Rephaims in Ashteroth Karnaim, and the Zuzims in Ham, and the Emims in Shaveh Kiriathaim,"

The Emim

The Emim were the giants who were dispossessed by the Moabites in Ar. Emim is a Moabite word for these people. The Moabites were the descendants of Lot through his older daughter. The Emim may have descended from a giant named Emi or Em. They dwelt on the other side of Jordan in Shaveh Kiriathaim, and were mostly destroyed by Chedorlaomer, King of Elam, but some were later destroyed by the Moabites.

Genesis 14:5; "And in the fourteenth year came Chedorlaomer, and the kings that were with him, and smote the Rephaims in Ashteroth Karnaim, and the Zuzims in Ham, and the Emims in Shaveh Kiriathaim,"

Deuteronomy 2:10-11; "The Emims dwelt therein in times past, a people great, and many, and tall, as the Anakims; Which also were accounted giants, as the Anakims; but the Moabites call them Emims."

The Zamzummim

The Zamzummim were a family of giants who were dispossessed from their land, which later was called the land of Ammon, by the Ammonites, the descendants of Lot through his younger daughter. The giant who started this family was possibly named Zamzumi or Zamzum. God destroyed these giants to make way for the children of Benammi, as He promised the daughter of Lot. *Genesis 19:38;* When it says they were destroyed by the LORD, it could mean a remnant of the Agent of Mass Destruction was responsible for their destruction. These giants were not killed by any men, and not by King Chedorlaomer.

Deuteronomy 2:20-21; "(That also was accounted a land of giants: giants dwelt therein in old time; and the Ammonites call them Zamzummims; A people great, and many, and tall, as the Anakims; but the LORD destroyed them before them; and they succeeded them, and dwelt in their stead:"

The Horim

The Horites, or Horim, were a family of giants descended from Hori, a giant. They dwelt in the area from Mount Seir to the wilderness of Paran. They were destroyed by Chedorlaomer, King of Elam, and those who escaped King Chedorlaomer were destroyed by the descendants of Esau and the Edomites, who dwelt in their stead.

Genesis 14:6; "And the Horites in their mount Seir, unto Elparan, which is by the wilderness."

Deuteronomy 2:12; "The Horims also dwelt in Seir beforetime; but the children of Esau succeeded them, when they had destroyed them from before them, and dwelt in their stead; as Israel did unto the land of his possession, which the LORD gave unto them."

Deuteronomy 2:22; "As he did to the children of Esau, which dwelt in Seir, when he destroyed the Horims from before them; and they succeeded them, and dwelt in their stead even unto this day:"

The Avim

The Avim could be descended from someone named Avi or Av. The Avim, or Avites, were a family of giants who dwelt in the land of Hazerim all the way to Azzah. They were destroyed by the Caphtorim. The Caphtorim were also giants. It appears they warred amoungst themselves.

Deuteronomy 2:23; "And the Avims which dwelt in Hazerim, even unto Azzah, the Caphtorims, which came forth out of Caphtor, destroyed them, and dwelt in their stead.)"

Joshua 13:3; "From Sihor, which is before Egypt, even unto the borders of Ekron northward, which is counted to the Canaanite: five lords of the Philistines; the Gazathites, and the Ashdothites, the Eshkalonites, the Gittites, and the Ekronites; also the Avites:"

II King 17:31; "And the Avites made Nibhaz and Tartak, and the Sepharvites burnt their children in fire to Adrammelech and Anammelech, the gods of Sepharvaim."

The Caphtorim

The Caphtorim were a family of giants who were descended from Caphtor. They destroyed the Avims, or Avites and took over their land. They, in turn, were dispossessed by the Philistines.

Deuteronomy 2:23; "And the Avims which dwelt in Hazerim, even unto Azzah, the Caphtorims, which came forth out of Caphtor, destroyed them, and dwelt in their stead."

Jeremiah 47:4; "Because of the day that cometh to spoil all the Philistines, and to cut off from Tyrus and Zidon every helper that remaineth: for the LORD will spoil the Philistines, the remnant of the country of Caphtor."

Amos 9:7; "Are ye not as children of the Ethiopians unto me, O children of Israel? saith the LORD. Have not I brought up Israel out of the land of Egypt? and the Philistines from Caphtor, and the Syrians from Kir?"

The Anakim

Anak was the father of a large and prominent family of giants known as the Anakim. Anak was the son of Arba, whose city, Kirjath-arba, was renamed Hebron by the Israelites. Anak sired three sons, also giants, named Ahiman, Sheshai and Talmai, who were driven out of Hebron by Caleb, and killed by the tribe of Judah as they possessed their inheritance.

The Anakim were the inhabitants of the Levant when Moses sent the spies up to spy out the land. It was the Anakim who generated the debilitating fear in the hearts of the wandering Israelites, which caused God to condemn them to forty years of wandering in the wilderness before they were allowed to enter their heritage. They were destroyed by Joshua and the Israelites when they took the land.

Numbers 13:22; "And they ascended by the south, and came unto Hebron; where Ahiman, Sheshai, and Talmai, the children of Anak, were. (Now Hebron was built seven years before Zoan in Egypt.)"

Numbers 13:28; "Nevertheless the people be strong that dwell in the land, and the cities are walled, and very great: and moreover we saw the children of Anak there."

Numbers 13:33; "And there we saw the giants, the sons of Anak, which come of the giants: and we were in our own sight as grasshoppers, and so we were in their sight."

Deuteronomy 1:28; "Whither shall we go up? our brethren have discouraged our heart, saying, The people is greater and taller than we; the cities are great and walled up to heaven; and moreover we have seen the sons of the Anakims there."

Deuteronomy 9:2; "A people great and tall, the children of the Anakims, whom thou knowest, and of whom thou hast heard say, Who can stand before the children of Anak!"

Joshua 11:21-22; *"And at that time came Joshua, and cut off the Anakims from the mountains, from Hebron, from Debir, from Anab, and from all the mountains of Judah, and from all the mountains of Israel: Joshua destroyed them utterly with their cities. There was none of the Anakims left in the land of the children of Israel: only in Gaza, in Gath, and in Ashdod, there remained."*

Joshua 14:12-15; *"Now therefore give me this mountain, whereof the LORD spake in that day; for thou heardest in that day how the Anakims were there, and that the cities were great and fenced: if so be the LORD will be with me, then I shall be able to drive them out, as the LORD said. And Joshua blessed him, and gave unto Caleb the son of Jephunneh Hebron for an inheritance. Hebron therefore became the inheritance of Caleb the son of Jephunneh the Kenezite unto this day, because that he wholly followed the LORD God of Israel. And the name of Hebron before was Kirjath-arba; which Arba was a great man among the Anakims. And the land had rest from war."*

Joshua 15:13-14; *"And unto Caleb the son of Jephunneh he gave a part among the children of Judah, according to the commandment of the LORD to Joshua, even the city of Arba the father of Anak, which city is Hebron. And Caleb drove thence the three sons of Anak, Sheshai, and Ahiman, and Talmai, the children of Anak."*

Joshua 21:11; *"And they gave them the city of Arba the father of Anak, which city is Hebron, in the hill country of Judah, with the suburbs thereof round about it."*

Judges 1:10; "And Judah went against the Canaanites that dwelt in Hebron: (now the name of Hebron before was Kirjath-arba:) and they slew Sheshai, and Ahiman, and Talmai."

Judges 1:20; "And they gave Hebron unto Caleb, as Moses said: and he expelled thence the three sons of Anak."

Some Infamous Giants

Og, King of Bashan

Sihon, King of the Amorites and Heshbon

Og, King of Bashan, was one of the most renowned of the children of the giants. He was destroyed by God before Moses and the Israelite advance. He was destroyed by a calamity, the same way King Sihon and his city Heshbon was destroyed. This is why they are grouped together in this thesis.

Numbers 21:33; "And they turned and went up by the way of Bashan: and Og the king of Bashan went out against them, he, and all his people, to the battle at Edrei."

Numbers 32:33; "And Moses gave unto them, even to the children of Gad, and to the children of Reuben, and unto half the tribe of Manasseh the son of Joseph, the kingdom of Sihon king of the Amorites, and the kingdom of Og king of Bashan, the land, with the cities thereof in the coasts, even the cities of the country round about."

Deuteronomy 1:4; "After he had slain Sihon the king of the Amorites, which dwelt in Heshbon, and Og the king of Bashan, which dwelt at Astaroth in Edrei:"

Deuteronomy 3:1-13; "Then we turned, and went up the way to Bashan: and Og the king of Bashan came out against us, he and all his people, to battle at Edrei. And the LORD said unto me, Fear him not: for I will deliver him, and all his people, and his land, into thy hand;and thou shalt do unto him as thou didst unto Sihon king of the Amorites, which dwelt at Heshbon. So the LORD our God delivered into our hands Og also, the king of Bashan, and all his people: and we smote him until none was left to him remaining. And we took all his cities at that time, there was not a city which we took not from them, threescore cities, all the region of Argob, the kingdom of Og in Bashan. All these cities were fenced with high walls, gates, and bars; beside unwalled towns a great many. And we utterly destroyed them, as we did unto Sihon king of Heshbon, utterly destroying the men, women, and children, of every city. But all the cattle, and the spoil of the cities, we took for a prey to ourselves. And we took at that time out of the hand of the two kings of the Amorites the land that was on this side Jordan, from the river of Arnon unto mount Hermon; (Which Hermon the Sidonians call Sirion; and the Amorites call it Shenir;) All the cities of the plain, and all Gilead, and all Bashan, unto Salchah and Edrei, cities of the kingdom of Og in Bashan. For only Og king of Bashan remained of the remnant of giants; behold, his bedstead was a bedstead of iron; is it not in Rabbath of the children of Ammon? nine cubits was the length thereof, and four cubits the breadth of it, after the cubit of a man. And this land, which we possessed at that time, from Aroer, which is by the river Arnon, and half mount Gilead, and the cities thereof, gave I unto the Reubenites and to the Gadites. And the rest of Gilead, and all Bashan, being the kingdom of Og, gave I unto the half tribe of Manasseh;

all the region of Argob, with all Bashan, which was called the land of giants."

Deuteronomy 4:47; "And they possessed his land, and the land of Og king of Bashan, two kings of the Amorites, which were on this side Jordan toward the sunrising;"

Deuteronomy 29:7; "And when ye came unto this place, Sihon the king of Heshbon, and Og the king of Bashan, came out against us unto battle, and we smote them:"

Deuteronomy 31:4; "And the LORD shall do unto them as he did to Sihon and to Og, kings of the Amorites, and unto the land of them, whom he destroyed."

Joshua 2:10; "For we have heard how the LORD dried up the water of the Red sea for you, when ye came out of Egypt; and what ye did unto the two kings of the Amorites, that were on the other side Jordan, Sihon and Og, whom ye utterly destroyed."

Joshua 9:10; "And all that he did to the two kings of the Amorites, that were beyond Jordan, to Sihon king of Heshbon, and to Og king of Bashan, which was at Ashtaroth."

Joshua 12:4; "And the coast of Og king of Bashan, which was of the remnant of the giants, that dwelt at Ashtaroth and at Edrei,"

Joshua 13:12; "All the kingdom of Og in Bashan, which reigned in Ashtaroth and in Edrei, who remained of the remnant of the giants: for these did Moses smite, and cast them out."

Joshua 13:30-31; "And their coast was from Mahanaim, all Bashan, all the kingdom of Og king of Bashan, and all the towns of Jair, which are in Bashan, threescore cities: And half Gilead, and Ashtaroth, and Edrei, cities of the kingdom of Og in Bashan, were pertaining unto the children of Machir the son of Manasseh, even to the one half of the children of Machir by their families."

I King 4:19; "Geber the son of Uri was in the country of Gilead, in the country of Sihon king of the Amorites, and of Og king of Bashan; and he was the only officer which was in the land."

Nehemiah 9:22; "Moreover thou gavest them kingdoms and nations, and didst divide them into corners: so they possessed the land of Sihon, and the land of the king of Heshbon, and the land of Og king of Bashan."

Psalm 135:11; "Sihon king of the Amorites, and Og king of Bashan, and all the kingdoms of Canaan:"

Psalm 136:20; "And Og the king of Bashan: for his mercy endureth for ever:"

Goliath

Goliath, was the most renown of all the giants mentioned in scripture, not as much for his life as for his death. He was a Gittite from Gath, he measured nine feet nine inches tall, and he was killed by young David in battle. After his death, Goliath's sword was kept at Nob, in the temple by Ahimelech the priest.

I Samuel 17:4; "And there went out a champion out of the camp of the Philistines, named Goliath, of Gath, whose height was six cubits and a span."

I Samuel 17:23; "And as he talked with them, behold, there came up the champion, the Philistine of Gath, Goliath by name, out of the armies of the

Philistines, and spake according to the same words: and David heard them."

I Samuel 21:9-10; "And the priest said, The sword of Goliath the Philistine, whom thou slewest in the valley of Elah, behold, it is here wrapped in a cloth behind the ephod: if thou wilt take that, take it: for there is no other save that here. And David said, There is none like that; give it me. And he inquired of the LORD for him, and gave him victuals, and gave him the sword of Goliath the Philistine."

Lahmi

Lahmi, the brother of Goliath was killed by Elhanan, the son of Jair.

II Samuel 21:19; "And there was again a battle in Gob with the Philistines, where Elhanan the son of Ja'are-oregim, a Bethlehemite, slew the brother of Goliath the Gittite, the staff of whose spear was like a weaver's beam."

I Chronicles 20:5; "And there was war again with the Philistines; and Elhanan the son of Jair slew Lahmi the brother of Goliath the Gittite, whose spear staff was like a weaver's beam."

Ishbibenob

Ishbibenob, a son of the giant of Gath, Goliath's father, was a giant who was killed by Abishai, the son of Zeruiah.

II Samuel 21:16-17; "And Ishbibenob, which was of the sons of the giant, the weight of whose spear weighed three hundred shekels of brass in weight, he being girded with a new sword, thought to have slain

David. But Abishai the son of Zeruiah succoured him, and smote the Philistine, and killed him. Then the men of David sware unto him, saying, Thou shalt go no more out with us to battle, that thou quench not the light of Israel."

Saph

Saph, or Sippai, the son of Goliath's father, was a giant who was killed by Sibbechai the Hushathite.

II Samuel 21:18; "And it came to pass after this, that there was again a battle with the Philistines at Gob: then Sibbechai the Hushathite slew Saph, which was of the sons of the giant."

I Chronicles 20:4; "And it came to pass after this, that there arose war at Gezer with the Philistines; at which time Sibbechai the Hushathite slew Sippai, that was of the children of the giant: and they were subdued."

The Unnamed Giant

This giant, the son of Goliath's father, carried six fingers on each hand, and six toes on each foot, and he was killed by Jonathan, the son of Shimeah who was the brother of King David.

II Samuel 21:20-21; "And there was yet a battle in Gath, where was a man of great stature, that had on every hand six fingers, and on every foot six toes, four and twenty in number; and he also was born to the giant. And when he defied Israel, Jonathan the son of Shimeah the brother of David slew him."

Il Chronicles 20:6; "And yet again there was war at Gath, where was a man of great stature, whose fingers and toes were four and twenty, six on each hand, and six on each foot: and he also was the son of the giant. But when he defied Israel, Jonathan the son of Shimea David's brother slew him."

More Unnamed Giants

The Four Giants Born to the Giant of Gath

These four may be referring to Ishbibenob, Saph, The giant with six fingers on each hand and six toes on each foot, and the brother of Goliath. And even if this reference is to four other sons of Gath, they were all brothers of Goliath in some form, at least they were all the sons of Goliath's father.

II Samuel 21:22; "These four were born to the giant in Gath, and fell by the hand of David, and by the hand of his servants."

I Chronicles 20:8; "These were born unto the giant in Gath; and they fell by the hand of David, and by the hand of his servants."

These are all the giants mentioned in the scriptures, by their families, by their names, by their physical characteristics, and by whom they were slain.

The impression these giants made upon men is evident by their prominence in scriptures. The lineage of these giants is almost as documented as is the lineage of God's own people. Other peoples of the time were not as meticulously documented.

The giants must have been truly admired by men.

Giants at the Second Coming

It is important to note the original giants were born when the "sons of God" came in unto the daughters of men, married them, and they sired children. This occurred in the days preceding the flood and even afterward, antecedent to it. *Genesis 6:4;*

Jesus said as it was in the days of Noah, it would be at his Second Coming. *Matthew 24:37-38;* He specifically mentions marrying and giving in marriage, as a precursor to his Second Coming, Remembering the events of those olden days, the "sons of God" married the daughters of men and the giants were the offspring. Since Jesus emphasizes the days of Noah in equating the times to the days of his Second Coming, he was probably referring to a recurrence of this giant's episode, in the last days.

Scientific Evidence

Scientific View

Scientists have identified two major different periods in prehistory showing evidence of massive destruction on the Earth by an outside celestial bombardment. The first, and most widely known and popular of the two was at 65 million years ago, when a very large asteroid or small comet or asteroid (scientists think about 10 kilometers {6 miles} in diameter) struck the Earth in the Gulf of Mexico on the edge of the Yucatan Peninsula near a small village named Chicxulub. The Chicxulub impact event left a crater in the side of the Yucatan Peninsula radiating out into the Gulf of Mexico measuring over 180 kilometers (110 miles) in diameter, all from a 10 kilometer (6 miles) wide chunk of rock.

This asteroid was the Agent of the extinction of the dinosaurs, and much of the flora and fauna on the Earth, approximately eighty five percent (85%) went to extinction.

The second, was an event of even greater destructive force at 240 million years ago. In this event ninety six percent (96%) of all plant and animal life on Earth was extinguished.

Biblical View

This is the event the Bible describes. *Genesis One;* The Reconstruction of the Earth was necessary because of this massively destructive event at 240 million years ago.

The Bible clearly tells us God destroyed "...*the world that then was...*" *II Peter 3:6;* "...*at his presence and by his fierce anger. Jeremiah 4:28;* Can it be known, in the physical realm, how He may have accomplished this feat?

God is quite capable of reaching into the natural realm, from the spiritual realm, to destroy the Earth, it does, after all, say He was present, *Jeremiah 4:26;* God operates within the realm of the physical laws whenever possible, laws established by Himself in the beginning, laws just as immutable as the spiritual laws, which He also established in the beginning.

This thesis proposes He did accomplish the destruction of the original Earth with a natural Agent. This thesis will attempt to show this physical Agent of Mass Destruction.

The Destruction of the Earth

At the date of 240 million years ago, a massive destruction befell the Earth.

Why?

The physical reason is known, the collision with the Agent of Mass Destruction.

But from a spiritual point of view, why?

If God created all this, why would He want to destroy it?

The only reason handed down to us causing God's fierce anger was the fall of Lucifer, to become Satan.

So this destruction was visited on Earth because of the fall of Satan. *Isaiah 14:20*;

The Shattered Earth

In the *First Chapter of Genesis;* between *Genesis 1:1; and Genesis 1:2;* there is a gap which takes the Earth from a pristine state to one of total chaos, as discussed above. This was accomplished by the impact event of 240 million years ago. The Bible says *"**the earth was without form and void**"*. The word *was, Hebrew: hayah:* הָיָה can also be, and has in many instances in the Bible, translated as *became* which gives a completely different meaning to the narrative and gives us a clearer understanding of the event, a catastrophe of monumental importance to the history of the planet, and literally an Earth shattering event. If the Earth *became* without form and void, then it must first have existed in a state which was ordered and ready for habitation, which is in accord with other scriptures. *Isaiah 45:18;*

The Deep

"... and darkness was upon the face of the deep. And the Spirit of God moved upon the face of the waters ..." Genesis 1:2;

Looking at a globe of the Earth and placing it with the Pacific Ocean centered in a frame of vision, it is evident the Pacific Ocean occupies one complete half of the Earth, or nearly so. This is because when the Earth again began rotating, gravity and centrifugal force acted upon the semi-sphere which the Earth resembled, and started the process of rounding the planet again. The flooded half underwater, after the dry land appeared on one side *"...**gathered together unto one place...**" Genesis 1:9;* rounded back out and formed the vast Pacific Ocean.

All planetary bodies above a certain size, in other words, able to hold their own gravitational field, whether moons or planets, tend naturally to mold into a sphere as they rotate on their own axis and revolve around their parents gravitational influence. The moons of Mars are two examples of moons not sufficiently large enough to hold their own gravitational field, even though they spin on their own axis and revolve around Mars, exerting a tidal influence on their parent, they are not spheroids. It doesn't matter if the composition of the planetary body is solid, liquid or gas, they all adhere to this spherical model, because the sphere is a result of the rotational spin, the revolutionary rotation and the gravitational field all acting in unison on the planet. It doesn't matter how fast or slow the rate of rotational spin, contrast the Moon, which rotates once every month, versus Jupiter which completes one rotation in less than nine hours, or the revolutionary rotation, contrast Mercury with Pluto. Gravitational fields do seem to have a great effect on the spherical nature of planets. On the other hand, none of the asteroids seem large enough to hold a gravitational field, and their rotational spins are erratic, their revolutions around the Sun, greatly interrupted by the planets they come into contact with, yet some of them are spheroid.

The Earth was no exception.

The *"deep"* in scripture is referring to the current Pacific Ocean. Keeping this in mind helps us to understand scriptural references to "the deep" in better context.

God is consistent in His treatment of words in the Bible, if a word (like the deep) means something in one place, it can and usually does, mean the same in every reference. However, there are many Hebrew words which have many nuanced meanings, as already shown (*Hebrew: hayah:* יָהָה). The deep (*Hebrew: tehom:* תְּהוֹם) is not one of these words, it means the deep in every usable instance.

When the celestial marauder came in contact with the Earth, it split the prehistoric Earth into three not quite equal pieces, leaving a gaping wound on the flat side of at least one of those pieces.

This gaping wound, left on the Earth, was referred to as *"the deep"* **Genesis 1:2;** in scripture. And the Earth was covered with water. **Genesis 1:2;** Water which later was separated from itself, above and below the firmament. **Genesis 1:7;** And later still, after the Earth again began its axial rotation, drained off the dry land, **Genesis 1:9;** filling this gaping wound, becoming the Pacific Ocean.

This left all the land mass on one side of the Earth, **Genesis 1:9;** with the waters filling the deep on the other side. This landmass is what is referred to as the continent of Gondwanaland. The sea on the other side scientists refer to as Panthalassa.

The Large Old Earth

The marauder hit the Old Earth, and cracked this larger planet Old Earth into three pieces.

The collision further pulverized the smaller of the three piece into numerous bits and fragments, which the marauder then deposited into their current orbit between Mars and Jupiter, along with the crushed remnants and fragmentary remains of its numerous other collisions with other members of our Solar System. These bits and pieces became the asteroid belt. This area of space, this place in the Solar System, could possibly be the place originally occupied by the larger prehistoric planet Earth.

It pushed the remaining largest piece containing about half the amount of the older larger Earth inward toward the Sun, becoming Earth as it exists today. Then the marauder, or what was left of it, pushed the middle sized piece into an orbit between the Earth and the Asteroid belt, becoming Mars, orbiting where it does in relation to the other members of the Solar System. There was much jockeying for position by members of the inner Solar System at this time, between 240 and 230 million years ago.

What was left of the Agent of Mass Destruction, and the interloper which caused so much havoc and left in its wake such widespread devastation, possibly did one of three things, and this is the mystery.

First, it could have exited the inner Solar System and went back out in its own orbit to the furthest reaches of the Solar System, its aphelion;

Second, it could have broken up and the pieces mostly filled the Asteroid Belt with the other broken bits remaining in the inner Solar System wreaking havoc among the planets, until there was little to nothing left except a few small NEOs (Near Earth Orbiting) asteroids causing trouble for the inner rocky planets.

Third, it may now be the planet Venus, having finally settled into its current orbit after millions of years of smashing into the other members of the Solar System. It probably started out as a much larger planet which was pared down to size from all these collisions and left much debris throughout the outer Solar System, (the Centaurs, etc.) and throughout the inner Solar System as well, (the NEOs).

If either of these last two options happened, which may seem likely, it would explain why scientists cannot find the planet they are looking for beyond the orbit of Pluto and Neptune. It could have caused the perturbations in Uranus' and Neptune's orbit millennia ago, when it encountered them, and then plunged into the inner Solar System where it established itself.

Pangaea

In *CE 1910*, American geologist Frank B. Taylor, CE *1860–1938,* proposed a theory of lateral (sideways) motion of continents causing mountain belts to form on their leading (front) edges. Building on this idea in *CE 1912*, German meteorologist Alfred Wegener, *CE 1880–1930*, proposed a theory known as Continental Drift. He proposed the continents moved, and were still moving, and were once part of one, large super-continent which he named Pangaea. Wegener was attempting to explain the origin of continents and oceans when he expanded upon Taylor's idea. His evidence included the shapes of continents, the physics of ocean crust, the distribution of fossils, and paleo-climatological data.

The Earth began to rotate on its axis again, perhaps millions of years after the original destruction. It was struck a glancing blow by a celestial body roaming through the inner Solar System. It struck the Earth in the "deep", and caused it to start spinning again. This roaming celestial body became Earth's Moon. It was probably originally a moon of the Agent of Mass Destruction, which was subsequently captured by the Earth. Before Charon, the prominent moon of Pluto, was discovered, the Earth/Moon ratio was the largest and most lopsided in the entire Solar System, and Charon may have also once been a moon of the Agent of Mass Destruction or of Saturn or Neptune.

This glancing blow caused the shattered half globe to begin its long voyage back to a spherical shape, a process still continuing today, a process called plate tectonics, or continental drift.

The Scientific Evidence for the Agent of Mass Destruction

System Wide Destruction

In the distant mists of Earth's prehistory, a planet sized celestial body came crashing down through the Solar System. Like a pin ball in an arcade game bouncing off the bumpers on the table, this destructive marauder was crashing off one celestial body after another, leaving in its wake a legacy of destruction and turmoil. When this malevolent intruder reached the Old Earth, it struck with ferocity, shattering the planet into pieces, destroying almost every living thing on the planet.

At the time of this event, the planets of the Solar System must have been aligned on one side of the Sun, so the marauder could affect the drama in succession. It started at the outermost planet, which was probably not Pluto, more likely Neptune, and in succession would have visited each of the planets in turn on its way to the Sun. It came careening through the Solar System, colliding with almost all the celestial bodies in the system, planets and moons, destroying some, and changing the shape of others with its impact. It was also changing the declination of axis and even the orbit of others, leaving in its wake a scene of cosmic devastation never before witnessed, or will likely ever be seen again.

The effects of this journey by this Agent of Mass Destruction can still be seen in our Solar System, and even on the Earth.

There is much evidence of this rogue planet throughout the Solar System.

To study the effects of this cosmic destroyer as it careened through the Solar System leaving chaos in its wake, see the chapter: The Agent of Mass Destruction and Its Effects on the Solar System.

To understand why this happened as it did, it is necessary to cast back in time and see if there was ever a time when the planets were all lined up on one side of the Sun, to accommodate this interloper in its destructive crusade.

Planetary Alignments

There were many times, in the history of the Solar System, when the planetary alignment was such to accomplish this feat. There are at least a few times every century when four or more of the planets all line up on the same side of the Sun. This occurrence has happened innumerable times in the history of the Solar System. Even the rare occurrence when all nine planets line up on the same side of the Sun has happened quite frequently in the history of the Solar System, every several thousand years or so. This is assuming a static Solar System which is not the case at all.

It is necessary to concentrate on those alignment events which are also accompanied by an event of mass destruction on the Earth to understand the alignment period in question occurred 240 million years ago, in the mid-Permian period, and was the one when the Agent of Mass Destruction made its fateful odyssey.

Gondwanaland vs. Pangaea

This study will now distinguish the difference between Gondwanaland and Pangaea, which were two different, though similar, landmasses in Earth's prehistory. Gondwanaland was the first landmass to emerge from an ocean covering the whole Earth at the beginning. It was the only landmass to emerge from the vast ocean, the first continent to appear from what scientists call Panthalassa, the worldwide ocean. It drifted around the globe in its early existence and moved over both cold spots (the poles) and hot spots (the equator), being mobile during its entire existence.

Then it started to break up.

During the Devonian Period, approximately 400 million years ago, the massive primordial landmass known as Gondwanaland broke into at least three landmasses. North America and Europe were together and sat near the equator, South America, Africa, Australia, India and Antarctica were one mass sitting at the South Pole, and much of Asia, including Siberia, was a single landmass located near the North Pole. Later, during the Permian, (about 260 million years ago) these landmasses once again migrated toward each other and fused into one landmass called Pangaea. The reason they fused again is uncertain to scientists. They have since again broken apart and through the process of continental drift, are formed into the landmasses of today. During the period when they were migrating to fuse into the continent of Pangaea, they were almost completely covered with water, if not entirely so, and scientists think this was a result of the end of an ice age and the influx of water into the ocean from the resulting ice melt.

I contend it was a much greater event, a destined collision with the Agent of Mass Destruction.

Hydrocarbons

Scientists think the vast reserves of hydrocarbons (oil) being harvested from the depths of the Earth in modern times was a result of the death, fossilization, and decay of the flora and fauna which have populated the Earth in the vast ages of the past. First hypothesized in *CE 1757* by the Russian scientist Mikhailo V. Lomonosov *CE 1711–1765* and eagerly propagated by the Swedish botanist Carl Linnaeus *CE 1707-1778*, the theory of the biological origin of hydrocarbons gained much momentum in scientific circles throughout the nineteenth and into the twentieth centuries. In *CE 1850*, when the German physicist Rudolf Clausius *CE 1822–1888* discovered the Second Law of Thermodynamics, this hypothesis came into question. It only gained a firm foothold again in the middle of the twentieth century when the use of "fossil fuels" (hydrocarbons), became more widespread, and Madison Avenue advertising executives needed a catchy phrase to sell their client's product, petrol (gasoline). Their commercials, showing the dinosaurs dying, liquefying and seeping into the depths of the Earth, popularized the idea.

Then, schools began teaching it as fact when, in reality, it is not fact, no more than the Theory of Evolution is fact.

Currently, *October CE 2011*, in Russia and the Ukraine, there are movements afoot to correct this misrepresentation. It is called the Russian-Ukrainian Theory of Deep, Abiotic Origins of Hydrocarbons. In this theory, scientists from Russia and the Ukraine are saying petroleum has an origin other than biology, meaning they did not come from the decayed bodies of the dinosaurs or the forests. Rather, they propose hydrocarbons are forced up to crustal levels from deep in the Earth by volcanic forces.

This Thesis proposes our deep reserves of hydrocarbons came from an extraterrestrial origin. This theory agrees with the research of Immanuel Velikovsky *CE 1895–1979*, a Russian born independent scholar, who wrote a number of controversial books on catastrophism and its effect on the planet during historical times, foremost of which were *"Worlds in Collision"* and *"Earth in Upheaval"*.

All this destruction was coming from the planet sized body striking the Earth in the far distant past, the Agent of Mass Destruction.

Creation

Science

In this section, it will be necessary to contrast biblical creationism with scientific creationism.

Science vs. Scriptures

In the beginning, the Scriptures state God created the Heavens and the Earth, and according to some hermeneuticists it was apparently about 6,000 years ago.

In the beginning, scientists are in general agreement, a Big Bang created everything about 13.8 billion years ago.

There is a tremendous dichotomy here.

Does one believe science, which bases its conclusion on interpreted empirical evidence, or does one believe God, the author of the Scriptures? If one believes science, it takes the exercise of our minds, using logic and intellect and an examination of the facts. If one believes God, it takes the exercise of our minds, using logic and intellect, and it also takes the exercise of our faith, because faith is the only way to please God, ***Hebrews 11:6***;, and because it is impossible to understand the things of the spirit with our logic and intellect, it takes faith to perceive these things.

The problem most people of intellect encounter when they are confronted with the Scriptures, and the inerrancy of same, is the Scriptures were interpreted by men. Men who have unfaithfully interpreted the Word of God, influenced by classic interpretations of prophesy. Interpretations causing the misunderstanding of God's Word in our world today. A misunderstanding compounded by the empirical evidence science is continually presenting us. Empirical evidence which stands on its own accord, unquestioned. Unquestioned because this evidence was methodically proven with the immovable laws of physics, chemistry, biology, mathematics and various other of the sciences. Sciences whose laws were set in motion by God Himself, at the creation, and whose laws are immutable. As immutable as the spiritual laws which were likewise set in motion by God at the creation, if not before.

A correct understanding of the scriptures is necessary to bridge the tremendous dichotomy between science and Scripture.

This essay will attempt to correct this understanding.

There is a way to believe both science and Scripture. The fault inherent in the belief structure of both science and scripture is the implacability of both adherents. Science must realize it doesn't have all the answers, and therefore the conclusions drawn about phenomena may come into question. Long held hypotheses of science (i.e. the Earth centered model of the cosmos and the flat Earth model) have been, and are continually being overturned by greater proofs and better theories (i.e. the Theory of Relativity). Likewise, interpreters of the Word of God must realize, with the multitude of unanswered questions concerning events past, present and future, in light of science and experience, their interpretations may also come into question, as has been shown in this work many times with different Hebrew renderings.

What is needed is a belief structure allowing the full scope of both science and scripture to meld and form one coherent hypothesis. Is this even possible?

Biblical Creationism

The Bible declares, ***Genesis 1:1; "In the beginning God created the heaven and the earth".***

This leaves no doubt as to the identity of the instrument of creation.

GOD.

God actually created three things during this initial act of creation. He created the Heaven, the abode of the spiritual beings (Heaven), the Earth, the abode of the physical beings (the Universe), and time (the beginning).

Scientists explain the beginning of the Universe happened in "The Big Bang", the Bible, however, clearly explains the motive force behind the creation was God. Scientists are at a loss to effectively explain the cause for the Big Bang, they merely know it occurred as evidenced by the background radiation being detected in every portion of the observable Universe. Scientists are fairly certain the Big Bang occurred, however, there seems to be no reason why it exploded, after being held in a compressed state for even a fraction of a second. The Bible fills in the missing pieces by revealing the reason for the Big Bang, and the reason is God. All things were created by the Word,

John 1:1-3; "In the beginning was the Word, and the Word was with God, and the Word was God. The same was in the beginning with God. All things were made by him; and without him was not any thing made that was made."

This further identifies the instrument of creation as Jesus. (The Word) Therefore this is proof Jesus is God. Jesus Christ was the manifestation of God to His creation, man. He chose to manifest Himself to us and accomplish our redemption through Jesus. Jesus is the physical manifestation of the God of all creation.

Ephesians 3:9; "And to make all men see what is the fellowship of the mystery, which from the beginning of the world hath been hid in God, who created all things by Jesus Christ."

It explains God accomplished the creation as Jesus, thereby confirming Jesus and God are the same, they are one. It was actually Jesus who accomplished all the acts of creation, and Jesus was God creating it all. It was only later He manifested Himself as Jesus, and allowed man to personalize God. All the acts of creation, and all the acts transpiring through the ages, from the creation, to the fall of Satan, to the fall of man, to the redemption of man by Jesus Christ, and through all of our history to the end of time were, are and will be ordained and ordered by God, as Jesus.

John 1:10; "He was in the world, and the world was made by him."

Jesus is the one who was in the world, who manifested Himself to us in our world although He was God, was the creator of all things. He was the one who established everything and upholds all things by His power.

John 1:14 "And the Word was made flesh, and dwelt among us, (and we beheld his glory, the glory as of the only begotten of our Father,) full of grace and truth."

This is another identification of the Word as Jesus. It clearly tells us He was made flesh, He took on the form of a man, His creation, and dwelt among us. He did this in order to accomplish the redemption of His fallen creation, man. God, as Jesus, established the law as a measuring rod to identify those who were redeemed from those who were not. Those who abided by the law, in totality, completely, (and no man ever accomplished this until Jesus lived) were able to redeem themselves. As previously stated, no man accomplished this. ***"For all have sinned and come short of*** *the glory of God." **Romans 3:23***; It was necessary for God Himself to redeem His creation, man. God did this as Jesus, the human manifestation of God in His own creation.

Hebrews 11:3; *"Through faith we understand that the worlds were framed by the word of God, so that things which are seen were not made of things which do appear."*

Everything existing today was made from nothing. This is in accord with what is known of the creation from scientific evidence. For science confirms, in the Big Bang Theory, before the creation, all matter was in a state of plasma, meaningless, compressed to the size of an atom. And all the atoms now making up all the matter in the Universe, visible and invisible (dark matter), were merely a collection of free flowing particles, compressed to an immeasurably small space, no bigger than an atom, plasma at best. The amount of pressure needed for this task is beyond anything imaginable, and surely beyond anything ever observed in the known Universe since. With God, all things are possible. **Matthew 19:26;** The pressure needed to hold the free flowing particles in this small space, and under this enormous pressure, is just a feeble hand squeeze to God.

Scientific Creationism

Adherents to biblical creationism, believe they know how the creation occurred, and they believe they know this because they have faith in what the Bible says about the event. Adherents to scientific creationism, however, have many theories from which to choose.

Creation, in the view of science, started about 13.8 billion years ago.

Three Theories of the Beginning

There are three main scientific theories on the creation of all things:
1. The Theory of the Eternal Universe.
2. The Big Bang Theory.
3. The Theory of Energy Conversion to Mass.

The Theory of the Eternal Universe has held the longest in man's history, but has pretty much fallen into disfavor today.

The Big Bang Theory, right now, is by far the most popular of the three.

The Theory of Energy Conversion to Mass is one using the Big Bang Theory as its basis. It adds to that theory to overcome the problems inherent in the Big Bang Theory, and where that theory is taking the Universe.

The Theory of the Eternal Universe

The Theory of the Eternal Universe basically maintains the Universe always was and always will be, just changing states from one age to the next, forever forming new stars from the stuff of the destroyed old ones. It is also called the Steady State Theory.

Anthropologists and archaeologists don't know much about the theories of the Universe held by the ancient dwellers on Earth, before the earliest civilizations emerged.

The earliest civilizations, the Sumerians, Assyrians and Babylonians, were very interested in the cosmos. They saw them more in the light of how they affected the lives of men on Earth than how they were formed or evolved. Their understanding of the Universe was intricately tied to their mythology and any understanding they possessed was seen through that lens.

The Theory of the Eternal Universe was the theory held by the learned men and scholars from the Greeks down through the Middle Ages. It just goes to show because a theory was dominant for a long time does not necessarily make it right (i.e. Aristotle's theory of gravitation, which simply stated the apple fell to the ground because it wanted to be on the ground).

Modern science has proven this theory wrong.

The theories held in the church should take a cue from this valuable lesson. For likewise, just because a theory or doctrine (dogma) is held for a long time does not necessarily make it right.

The Big Bang Theory

The Big Bang Theory, the theory most popular at present is the theory currently being taught in school and is the theory held by most scientists today.

Basically it proposes, before the beginning, all matter in the Universe was compressed to the size of an atom, and there was no cohesion to the free flowing particles (plasma), there were no atoms intact at this point, just free flowing particles, in essence, it was nothing.

At the moment of creation, this plasma exploded outward to form the stars and galaxies and everything existing today. It further asserts the Universe is still expanding outward.

So, in essence, everything was made from nothing.

This outward expansion may continue forever until all stars burn out, or it may, at some future point, start contracting in what is being called the Big Crunch.

The only problem is there is not enough mass currently available in the Universe to allow for this Big Crunch.

Scientists today are searching the Universe for more matter to allow for the Big Crunch. They postulate the necessary matter may be in three possible conditions.

First, it may be in the form of dark matter.

Second, it may be in the form of dark energy.

Third, it may be found in Black Holes.

Without this missing matter, there will be no Big Crunch.

They have concluded, therefore, the Big Freeze, where expansion continues to the point where all stars simply burn out, leaving a dead black Universe, could be the destiny of the Universe.

This thesis proposes a solution to their dilemma.

The Theory of Energy Conversion to Mass

The Theory of Energy Conversion to Mass basically expostulates all matter, space and time were originally in the form of energy (plasma) compressed to the size of an atom, similar to the Big Bang Theory. As the space, time and matter exploded outward in the Big Bang, most of the mass stayed in the form of energy and slowly converts to mass over billions of years. This mass, in the form of energy (dark energy), is travelling at the speed of light squared, which means it is unobservable.

In the star forming regions, there are tremendous gravitational fields acting upon this energy, slowing it down imperceptibly, just enough to bring it below the speed of light squared, a speed necessary to maintain mass in the form of energy. Energy is becoming mass, as stated in Einstein's Special Theory of Relativity ($E=mc^2$). As this energy converts to mass (makes stars), the mass adds to the gravitational field already present, which acts upon more of the free flowing energy, thereby causing more energy to slow and convert to mass, thereby creating more gravitation, etc., etc., etc.

One of the main problems with the Big Bang Theory is there is not enough mass present as a result of the Big Bang to allow for a Big Crunch.

This theory, The Theory of Energy Conversion to Mass, which is the author's own, posits the mass needed in this Universe to effect the Big Crunch, which God in His Word declares will occur, is still in the form of energy.

This theory derives from Einstein's Special Theory of Relativity, ($E=mc^2$), first expressed by the German/American physicist Albert Einstein. *CE 1879-1955* Einstein's theory *CE 1905* theorizes mass, traveling at the speed of light squared, will convert to energy. If an atom of mass travels at the speed of light squared, Relativity Speed (186,000 x 186,000 miles per second, or 34,596,000,000 miles per second) it will convert to energy and stay in this state as long as it travels at this speed. Energy is merely mass traveling at a tremendous speed, Relativity Speed.

Conversely, energy, which slows below the speed of light squared, will convert to mass. As energy, mass has no substance. If it slows, it takes on substance and becomes an atom of mass again (hydrogen). All the physical Universe is made up of atoms of mass.

Scientists used Einstein's Special Theory of Relativity to develop the Atomic Bomb by fission of Uranium, turning mass to energy according to the Theory of Relativity.

This Theory turns it around and proposes energy converting to mass.

In the star forming regions of the Universe, like in Orion, this process is taking place, as the gravitational fields there are slowing energy down to form mass. When enough energy converts to mass in the Universe, and creates a sufficient amount of mass to allow a Big Crunch, then all the matter in the Universe will begin an implosion. This process will proceed considerably faster than the expansion, speeding the matter back to the speed of light squared in a relatively short space of time, thereby returning the mass to a state of energy, compressing it again to the size of an atom, converting all matter back to a state of plasma and proceeding to the next event (the next Big Bang and the next Universe).

Currently, according to scientific measurements, there is only about five percent (5%) of the mass needed to effect a Big Crunch. This has baffled scientists because they estimate there needs to be approximately seventy percent (70%) matter to allow the universal contraction. Currently, the Universe is expanding at an increasing rate, which means when there is enough mass present, the Universe will contract at a steadily decreasing rate until it reaches the point where it will then stop expanding and start to contract and that contraction will be at an ever increasing rate until it becomes just a plasma again, held into the size of an atom. Then will come another Big Bang, etc.

Scientists theorize dark matter and dark energy to account for the missing mass. The Theory of Energy Conversion to Mass can allow this matter to still reside in the Universe as energy, and would indicate man is living in a very young Universe indeed. If it took 13.8 billion years to accumulate five percent (5%) of needed mass, then at a steady rate, the Universe could age to 276 billion years. There is evidence the Universe is expanding at an increasing rate, and therefore one can theorize an increasing rate for the contraction as well. Suffice it to say, if this theory is true, the Universe has at least 150 billion years longer before the Big Crunch, and the next Universe.

Then all the matter in the Big Crunch, will once again be in a state of plasma, (nothingness) contained in the size of an atom.

This has happened in the past, at this Universe's Big Bang, and will happen again in the future to unveil the next Universe. How many times in the dateless past this has occurred, or how many times in the ageless future this will occur is unknowable on this side of the veil (death). It can be conjectured to have happened at least once before, and will happen at least once more, based on scripture.

The Agent of Mass Destruction and Its Effects on the Solar System

The Solar System, which man calls home, is an amazing storehouse of evidence demanding a verdict. All through the Solar System there are mysteries and anomalies defying conventional wisdom, speaking most clearly of a catastrophic history in the systems formation and history.

There is evidence of catastrophe and destruction in every single planet of this system, from its outer reaches to the Sun itself, the motor driving it all. This thesis will investigate this evidence of destruction from the outer edges of our system to the Sun, the direction the Agent of Mass Destruction traveled.

The Solar System is a vast area, occupying a small corner in an outer wing of the even more immense Milky Way Galaxy. There are many components of the Solar System, the most important being the planets and the Sun itself, but there are also other elements composing much of the mass of the Solar System. There is the Oort Cloud, the Kuiper Belt, The Asteroid Belt, the moons, the rings, the comets, the meteorites and all the dust and gas filling the spaces in between. This thesis proposes, each of these elements, and all of them inclusively, show evidence of the Agent of Mass Destruction.

This thesis is going to discuss these elements of the Solar System to show the evidence present which unequivocally proves the theory herein presented, a theory of catastrophe, a theory of destruction, a theory of extinctions and a theory of celestial reshaping. These theories are not new, they were present in our Mythology and in our religious texts for millennia. It is time to uncover them anew for the coming millennia.

The Minor Bodies

The Oort Cloud

The first of these elements, lying at the very edge of the Sun's influence, is the Oort Cloud, named in honor of its proposer, the Dutch astronomer Jan Hendrik Oort, *CE 1900-1992*. Even though there is no solid proof yet, for the existence of the Oort Cloud, there is sufficient evidence the many clumps if ice which come to us as comets of long range must reside somewhere beyond the far reaches of our Solar System.

The Oort Cloud is a theoretical assemblage of these many millions of ice clumps of varying size supposedly residing at the very fringes of the Solar System and make up the place of origin of the comets periodically careening through the planetary realm in a headlong plunge toward the Sun and out again.

It is theorized other stars sometimes come close enough to the Oort Cloud to dislodge some of these icy clumps and send them hurtling inward toward the Sun. As they approach the Sun, the gases from the solar wind cause the icy particles in their composition to heat up, fluoresce and burn away, giving the comets their tails. As they round the Sun and head back into the further reaches of the Solar System again, they lose their tail and freeze again. Some comets of short range have been making this elliptical odyssey for millennia and some have been catalogued by astronomers.

The Solar System, in its revolution around the center of the Milky Way Galaxy, moves up and down in its movement through the galactic disc. Looking at the revolution of the Solar System from the side of the Milky Way Galaxy, one would see it revolving with a remarkable wobble moving above and below the galactic disc. As it makes this wobbly journey, it almost assuredly passes through cold and hot spots within the plane of the ecliptic of the Milky Way Galaxy, which may account for ice ages and greenhouse effects throughout Earth's pre-history. When it is above, it is higher than the plane of the disc and when it is below it is lower than the plane of the disc. Every time it passes from below to above or from above to below, it must pass through the plane of the ecliptic, and during these times, the Solar System makes contact with other stars. These contacts with other stars may disrupt bodies of ice and rock residing in the Oort Cloud.

These disrupted pieces of ice and rock become the comets. The comets of long range sweep around the Sun and return to the outer reaches, whereas the comets of short range begin out there, but get captured by the planetary gravitational influences of the major planets and end up in short range orbits mostly defined by the orbits of those major planets around the Sun.

Whereas scientists surmise a passing star dislodges these ice clumps and sends them hurtling in toward the Sun, it would be just as legitimate to assume the Agent of Mass Destruction is the dislodging instrument. In one of its sweeps to the outer edges of its long and extremely elliptical orbit, it may indeed come into contact with the place of residence of these icy clumps, the Oort Cloud, and affect their orbits to send them hurtling in toward the Sun.

And this planet of very long duration and very oblate orbit may be what scientists are now looking for beyond the reaches of the orbits of Neptune and Pluto. A planet the scientists are calling Nemesis, or Planet X.

Nemesis

In *CE 1888*, when the planet Uranus was first discovered by the German, British astronomer William Herschel *CE 1738-1822*, scientists immediately started looking for another planet further out. The reason for this further exploration was because of a perturbation in the orbit of Uranus signifying another large body somewhere out beyond the orbit of Uranus. With exact calculations on the perturbation of the orbit of Uranus, the scientists in short order found Neptune, and were then smugly satisfied they completed the extent of the Solar System's family.

Then Herschel noted some other large perturbations in the orbit of Neptune, as well as some other unaccounted for smaller perturbations in the orbit of Uranus not answered by the presence of Neptune, His calculations of the position and orbit of Neptune indicated there would be another planet even further out still.

Immediately scientists began their search for this new planet, thinking they would find it in much the same way they found Neptune. When Pluto was discovered, they again were feeling satisfied, but the size of Pluto was totally insufficient to account for the large perturbation in Neptune's orbit and the anomalous orbit of Pluto only added to the enigma.

Today, scientists are still searching for a larger planet, further out than Neptune and Pluto. There are still perturbations in the orbit of Uranus and Neptune yet to be answered by the presence of Pluto, and they know there is another extremely large planet out there somewhere, with a rather large specific mass. It may even be a rocky planet, although most of the outer planets in the outer Solar System are gaseous in nature, Pluto being an exception. Pluto was most probably a rejected moon of Saturn, Uranus or Neptune (most probably Neptune since the orbits of these two planets intersect) exiled to the outer reaches of the Solar System at some undetermined time in the distant past.

In the far distant reaches of our Solar System there almost certainly is another planet. This planet has yet to be discovered, but due to the erratic orbits of Neptune and Uranus, it must be there. The size and relative mass of the Pluto/Charon duo is insufficient to cause the perturbations of orbit measurable in Neptune and Uranus indicating this other planet is out there, somewhere.

Some scientists have already named this planet Nemesis, or Planet X (for 10) since they are also naming an unfound dark star Nemesis. This work names it the Agent of Mass Destruction.

If this planet, in a wildly elliptical orbit taking it to the extreme edges of the outer Solar System and back in very close to the Sun, perhaps even inside the orbits of Earth or Venus, is the Agent of Mass Destruction discussed in this study, then the name is appropriate. In Greek Mythology Nemesis was the goddess of vengeance and retribution.

An interesting thing about the naming of the planets and other celestial bodies in the Solar System, the name assigned to each from Greek and Roman mythology, is appropriate as a representation of the possible role each one played, and continues to play, in the formation and shaping of the Solar System, whether protagonist or antagonist, in the drama playing out physically in the Solar System's formative years. For instance, Jupiter is the largest planet and has the most influence on all the other planetary bodies and other loose debris flying throughout the system. Jupiter was named for the king of all the gods in the Roman pantheon.

And yet, other bodies in the Solar System were just named for their discoverers, like the Kuiper Belt.

The Kuiper Belt

The Kuiper Belt, named after its discoverer, Dutch, American astronomer Gerard Kuiper *CE 1905-1973*, is an area outside the orbit of Neptune and Pluto, where numerous asteroids and comets and chunks of rocky ice reside in orbit around the Sun. Comets with periods of less than 500 years are said to reside in the Kuiper Belt, whereas comets of long period are said to reside in the Oort Cloud way out on the very fringes of the system, where the influence of the other members of the system is the weakest.

The fact there are comets residing in the Kuiper Belt is further evidence of an outside influence on these particular comets, to allow them to move from long period to short period comets. This influence could not have come from any planets inside its own orbit, like Saturn or Jupiter, or else they would reside within those orbits around the Sun. The Kuiper Belt orbits the Sun outside the influence of the outer planets normally responsible for changing a comet from long period to one of short period, usually Jupiter or Saturn, sometimes Uranus and Neptune. This speaks to an external influence orbiting outside the orbit of Neptune, where the Kuiper Belt orbits the Sun, and this is prime evidence for the Agent of Mass Destruction, the planet scientists are tentatively calling Nemesis, the planet they are diligently searching for at present, and have been since Neptune was discovered. They may not find the missing planet out there, for this bold thesis avers it is the Agent of Mass Destruction and currently may be hiding in plain sight. More on that later.

The Centaurs

The Centaurs are a group of wayward comets or asteroids with irregular orbits around the Sun in the spaces near the outer gas giants, especially between Saturn and Uranus. There is estimated to be close to forty four thousand Centaurs in this area of the Solar System.

The most well studied of the Centaurs is 2060 Chiron, a minor planet first observed in *CE 1999* by German astronomer Hans Scholl, born *CE 1942,* (not to be confused with Charon, Pluto's moon). Chiron measures roughly 320 kilometers (200 miles) in diameter and orbits the Sun every fifty one Earth years (one Chironian year). It has the characteristics of a comet but is much larger than comets usually attain, making scientists think it might be an asteroid. If it is an asteroid, it is the first one with a known atmosphere and rings, which have just recently been detected.

The Centaurs are believed to have once resided in the Kuiper Belt, but have become captured by the gravitational influences of the outer gaseous planets in their revolutions inward to round the Sun. They did not have enough momentum to escape to the outer realm of the Solar System, back to the Kuiper Belt, and they are now residents of the area of the two gas giants mentioned earlier. Although scientists are fairly certain the stronger gravitational pull of the gas giants is responsible for the Centaurs, it is entirely possible the Centaurs are the scattered remains of an enormous encounter between the outer planets, and even the ancient Earth, with the Agent of Mass Destruction. As the marauder began its trek back to the farther reaches of its orbit, after it shattered the Old Earth, it deposited much debris of rock and ice all through the Solar System, and the gas giants merely herded these pieces into their present orbits, like the shepherd moons of Saturn keep the rings intact.

The Asteroid Belt

The Asteroid Belt, which this thesis proposes are remnant pieces of the Old Earth, were left in their present orbit after this wandering ravisher broke the Old Earth into three major pieces, the smaller third of which was further shattered into numerous pieces which are now the asteroids, and pushed the largest piece of the Old Earth closer toward the Sun to become the planet the human race now inhabits, the present Earth. The second large piece was pushed to another orbit, where it became Mars. The planet Earth, the planet Mars, and the numerous asteroids in the inner Solar System were once a much larger rocky planet orbiting the Sun between Mercury and Jupiter. This thesis also conjectures Venus was not quite yet in its present orbit. More on this later.

Some of the Asteroids are quite large, about two hundred are more than 97 kilometers (60 miles) in diameter, whereas thousands of others are mere dirt clods. They revolve around the Sun in an orbit between Mars and Jupiter, however there are two groups of asteroids called Trojan asteroids leading and trailing Jupiter in orbit by sixty degrees (60°). There is also a group of about seventy five asteroids crossing the orbit of Mars, known as the Amor asteroids, and another fifty crossing the orbit of Earth called the Apollo asteroids, and another group of less than ten travelling inside Earth's orbit, crossing the orbit of Venus, called the Aten asteroids. In *CE 1977* an asteroid was observed between the orbit of Saturn and Uranus called 2060 Chiron. There are many observed there and are now referred to as the Centaurs as described above.

The fact the asteroids are spread out throughout the Solar System is further proof of the Agent of Mass Destruction. As this nomadic planet came careening down through the Solar System in the distant past, shattering many worlds, it spread the debris throughout the areas where it traversed. In the outer Solar System, where the gas giants reside, it left the debris of chunks of rock and ice, and closer in to the inner planets, it left the rocky asteroids. Since the asteroids and the Earth and Mars are related in composition, this thesis proposes they were once all parts of a much larger planet, a planet hit and shattered by the Agent of Mass Destruction. The largest piece being the Earth, the middle sized piece becoming Mars, and the smaller piece being broken up further and becoming the asteroids in the Asteroid Belt orbiting between Mars and Jupiter.

There are too many asteroids to explore them in depth here. It is only necessary to analyze a sampling to see there is indeed much evidence in this area of space of this catastrophe, the calamitous romp through our Solar System by the Agent of Mass Destruction.

Some Prominent Asteroids

There are millions of asteroids in the Asteroid Belt, from as large as Ceres to as small as mere dust particles. Those prominent enough to deserve a name also have a designation number according to their order of discovery.

1 CERES – Ceres is classed a minor planet, though it resides in the asteroid belt. It is 945 kilometers (587 miles) in diameter and is spherical. It rotates on its own axis every nine hours and four minutes (one Ceresian day). Ceres is the only body within the Asteroid Belt known to be spherical based on its own gravity. Ceres was discovered in *CE 1801* by Italian astronomer Giuseppi Piazzi *CE 1746-1825*. Ceres was named after the Roman god of agriculture.

4 VESTA - Vesta is 525 kilometers (326 miles) in diameter and spheroid in shape. It makes one rotation on its own axis every five hours and twenty minutes, Earth time (one Vestan day). It is smaller than the state of Arizona. Vesta was discovered in *CE 1807* by German astronomer Heinrich Wilhelm Olbers *CE 1758-1840*. Vesta is named for the Roman goddess of hearth and home.

2 PALLAS – Pallas is slightly smaller than Vesta at 512 kilometers (318 miles) in diameter. Pallas is more an ellipsoid than a spheroid and has irregular rotation, it wobbles. Pallas was also discovered in *CE 1802* by German astronomer Heinrich Wilhelm Olbers. Pallas is named after the Greek goddess Pallas Athena.

3 JUNO – Juno measures 320 kilometers (199 miles) in diameter and orbits the Sun in four years and three months (one Junoan year). Juno was discovered in *CE 1804* by German astronomer Karl L. Harding *CE 1765-1834*. Juno was named after the main Roman goddess, Juno.

10 HYGIEA – Hygiea is smaller than Pallas with a diameter of 350-500 kilometers (220-310 miles). Although larger than some other asteroids which were discovered before it, it was founded in *CE 1849* by Italian astronomer Annibale de Gasparis *CE 1819-1892*. It is relatively darker than most asteroids and has a highly irregular shape. Hygiea was named after the Greek goddess of health.

243 IDA AND DACTYL - Ida is irregular in shape measuring 56 by 24 by 21 kilometers (35 by 15 by 13 miles) in circumference. It rotates on its own axis in just over four hours and thirty minutes (one Idan day). A flyby of Ida in *CE 1993* by the Galileo spacecraft discovered Ida supported a moon of its own, the first Asteroid discovered with one. The moon is named Dactyl and measures about 1.5 kilometers (1 mile) across. Ida was first discovered in *CE 1884* by famous Austrian astronomer Johann Palisa *CE 1848-1925*. In Greek mythology, Ida was a Cretan nymph who raised Zeus as a boy.

253 MATHILDE - Mathilde is 52 kilometers (33 miles) in diameter and completes one orbit in four years and three months (one Mathildean year). Mathilde was also discovered by Johann Palisa in *CE 1885*. Mathilde was named after the wife of the Director of the Paris Observatory, Maurice Loewy *CE 1833-1907*.

951 GASPRA - Gaspra is approximately 20 kilometers (12.5 miles) across. It rotates on its own axis in just over seven Earth hours (one Gaspran day). It is very irregular in shape and was found in *CE 1916* by Russian astronomer G.N. Neujmin *CE 1886-1946*. Gaspra was named after a Black Sea resort which was popularly visited by many of Neujmin's friends like Maxim Gorky *CE 1868-1936* and Leo Tolstoy *CE 1828-1910*.

1620 GEOGRAPHOS - Geographos is 5.1 by 1.8 kilometers (3.2 by 1.2 miles) in circumference. It takes just one year three months for Geographos to make an orbit around the Sun (one Geographosian year). It was discovered in *CE 1951* by American astronomers Albert George Wilson *CE 1918-2012* and Rudolph Minkowski *CE 1895-1976*. Geographos was named in honor of the National Geographic Society.

4179 TOUTATIS – Toutatis is similar to Castalia in appearance. The asteroid seems to be composed of two segments in contact with each other. The larger side is 4 kilometers (2.5 miles) across and the smaller side is 2.5 kilometers (1.6 miles) across. It makes an orbit in four Earth years (one Toutatisian year). It was first discovered in *CE 1934,* then lost shortly thereafter. It was rediscovered in

CE 1989 by the French astronomer Christian Pollas, born *CE 1947,* and named Toutatis at this rediscovery for the Celtic god of protection.

4769 CASTALIA - Castalia is 1.8 kilometers (1.1 miles) in width and has a dumbbell shape, it is conjectured two asteroids fused together during a collision. It is an Apollo class asteroid (meaning it crosses the orbits of Mars and Earth and it measures 75 kilometers (47 miles) across lengthwise. It completes one orbit around the Sun in a little over one Earth year (one Castalian year). It was found on photographic plates in *CE 1989* by American astronomer Eleanor F Helin *CE 1932-2009* from Caltech. It was named after Castalia, a nymph in Greek mythology.

9969 BRAILLE - Braille is 2.2 kilometers (1.3 miles) at its widest and 1 kilometers (.6 miles) at its narrowest. It is a Mars crossing asteroid and it makes one revolution around the Sun in a little over three and a half Earth years (one Braillean year). Braille resembles Vesta very closely and scientists think Braille was once a part of Vesta, evidence of past collisions. Braille was discovered in *CE 1992* by American astronomers E.F. Hein *CE 1932-2009* and K.J. Lawrence, born *CE 1964*, from Palomar Observatory. It was named after the inventor of the writing system for the blind, Louis Braille *CE 1809-1852.*

There are many more asteroids which is beyond the scope of this study.

Castalia, Toutatis and Ida and its moon give possible evidence for the Agent of Mass Destruction

Vulcan

In the late *CE 1800's*, French mathematician Urbain le Verrier *CE 1811-1877* observed a small planet orbiting very close to the Sun, which they named Vulcan. These observations were made over a two year period, in all parts of the world and from numerous observatories on Earth, and then the planet completely disappeared from view, never to be seen again. There are some skeptical scientists who believe nothing of substance was observed and all those many astronomers, from numerous observatories all over the world, were simply mistaken. Those who believe some actual celestial body had been observed and correctly documented are fairly certain the planetoid they named Vulcan came too close to the Sun and was drawn in and totally burned up, and there was much official speculation about its origin and history at the time. Since then, many asteroids have been discovered between the orbits of Mercury and the Sun and these are now designated as the Vulcanoids. Some have been in their orbits for centuries, since their first discovery, and others have been observed falling into the Sun and being obliterated, precedent for the demise of Vulcan.

This study contends this planetoid was a small planet like relic of the collision herein discussed, possibly even a piece of the shattered Earth, or a dislodged moon from the outer Solar System having been dragged along behind the Agent of Mass Destruction from the course of its journey in the far distant past. However, observations were too brief and inconclusive to form any solid conclusions about this wayward member of the society of planets.

Nemesis, The Death Star

Along with the mysterious and undetectable planet being eagerly sought among modern astronomers named Nemesis and sometimes called Planet X, there is also a supposedly mysterious and undetectable star, also named

Nemesis, being sought out in the vicinity of the outer reaches of the Sun's influence, a binary star to the Sun. The reason the unknown planet is sometimes called Planet X is to alleviate confusion with this unidentified star.

The reason these scientists propose this supposed binary star to the Sun is because most stars in the galaxy, of the yellow star variety, travel in pairs, our Sun seems to be an exception.

Every 28 million years, or so, Nemesis supposedly approaches close enough to the Solar System to disturb groups of comets orbiting around the Solar System outside the planets. The comets then bombard the inner Solar System, including Earth.

There is not a shred of evidence for the existence of a star named Nemesis, however. Since it has not been found, its supporters say it must be small and dark. It is, they appropriately say, unfortunately (or conveniently) at the farthest end of its elliptical orbit now (aphelion), too far away for easy detection.

Even if a Death Star once existed, its orbit would be distorted by close passages to the Sun and any periodicity would have disappear after a few return visits. Even if the extinctions are periodic, Nemesis could not explain them.

Altogether, the Nemesis idea, though vivid, was never satisfactory, it didn't prevent it from spawning tremendous publicity, several TV programs, two covers of Time, and at least five books!

Planet X

Problems with the Nemesis idea led almost immediately to the Planet X hypothesis.

Planet X is a hypothetical planet supposedly orbiting somewhere outside the orbit of Neptune, maybe even beyond the Kuiper Belt. If Nemesis, the Death Star, cannot produce regular comet showers, perhaps Planet X, the Death Planet, could shower the inner Solar System with comets.

Planet X necessarily must have substantial mass (much more than Earth), yet the Infrared Astronomical Satellite (IRAS), an infrared orbiting telescope, has, as yet, proven unable to find it. In any case, computer models show Planet X cannot produce periodic showers of comets any more than Nemesis.

So, star or planet, the continuous inflow of comets from the outer reaches of the Solar System has another cause.

Planet X is basically a smaller conjectured version of the star Nemesis described above.

Summation of the Minor Bodies

This concludes all the descriptions of the loose pieces and meandering objects of the Solar System, and along with the planets and moons, this is pretty much a description of the known Solar System as understood today. Now it is time to explore the planets and moons to uncover more evidence of the Agent of Mass Destruction.

The Major Bodies

The Sun

Even the Sun shows revealing evidence of this cosmic upheaval. In the *CE 1600's*, there was a period of over seventy years when the regular cycle of sunspot activity, which occurs every eleven years, was interrupted. During this time, which is known as the Maunder Minimum, named for the famed British solar astronomers, Annie Russell Maunder *CE 1868–1947* and Edward Walter Maunder *CE 1851–1928*. From *CE 1645-1715* there was a total absence of sunspot activity. Scientists are at a loss to explain this event, this Maunder Minimum, to their own satisfaction. They are unsure of the principle cause of the event, or its ultimate implications for the Solar System, but they are definitely sure it happened. There is evidence on the Earth, found with dendrochronology, the reading of tree rings, of the Maunder Minimum. There was no sunspot activity during this seven decade span. Scientists try to explain it as triggered by stimuli within the Sun itself, though they are unable to explain the nature of this Sun shaking stimulus.

This thesis proposes in the *CE 1600's*, a leftover piece of the fateful collision of the Old Earth with the Agent of Mass Destruction, a piece of immense size, crashed headlong into the Sun and caused a slight cooling of the surface of the Sun, thereby precipitating the Maunder Minimum. This piece was probably much larger than Vulcan, for when Vulcan crashed into the Sun, in the *CE 1800's*, there was no suspension of sunspot activity, in fact, there seemed to be no adverse effects on the Sun from Vulcan, but Vulcan was quite small and innocuous. The suspension of sunspot activity in the *CE 1600's* was undoubtedly caused by an extremely large piece of remnant rock indeed.

The Sun has acted as a huge vacuum cleaner for the inner Solar System for most of its history, just as Jupiter has for the outer Solar System. The interloper, in its ancient trek down through the entire Solar System deposited much debris and wreckage for the mighty Sun to vacuum and dispose of in its own inimitable way.

Neutrinos

There may be other evidence of the Agent of Mass Destruction evident on the Sun. It is hard to tell, for the Sun has an ever changing surface. There was, however, an interesting experiment conducted in the early *CE 1970's* in a deep salt mine which was no longer of any use as a mine in South Dakota, USA.

They (the scientists) filled a large tank deep in the mine with chlorine water. The purpose of the experiment was to measure the number of neutrinos being emitted by the Sun.

When a neutrino strikes a chlorine atom, it converts it into a radioactive argon-7 atom, the experiment was intending to measure the number of argon-7 atoms per cubic centimeter to determine the relative age of the Sun. The number of converted neutrinos were expected to be quite high, and as the Sun ages, the rate of neutrino release dramatically lessens, until the Sun itself swells and expands to the red giant stage emitting no more neutrinos, and dies by collapsing into a white dwarf star. Since the Sun is only four and a half billion years old, (as scientists surmise) and, like most yellow stars of its class, it has an expected life of between twenty to thirty billion years, they expected quite a large number of neutrinos per cubic centimeter, therefore the need for the experiment to measure them..

Neutrinos pass through ordinary matter leaving no discernable evidence of its transit. There are neutrinos passing through each person on Earth every day. The reason they put the vat deep in the mine was to filter out other reactive agents present in the atmosphere, since only neutrinos travel untrammeled through matter, having it deep in the mine was no hindrance to the neutrinos passing through.

The result was not what the scientists expected. They expected there to be between one hundred to two hundred neutrinos per cubic centimeter, (forty to eighty per cubic inch) however, there were only about five to six neutrinos per cubic centimeter (two to two and a half per cubic inch).

The scientists concluded their experimental control was based on a wrong premise and they determined the experiment was null and void.

This thesis contends their control was not at all wrong, and the experiment did accurately measure the age of the Sun and accurately gave a true measure of the remainder of its life. The remaining life of the Sun is not as long as previously thought. It is a dying Sun, and God has set it so. It won't be ten to fifteen billion more years before this Earth is melted with fervent heat, **II Peter 3:10**; rather much less time is remaining, perhaps only a few millions of years instead of billions, or maybe even just thousands.

The melting of this Earth with fervent heat is an exact description of the Sun going to the red giant stage.

The Sun is a tremendous ball of fusion created nuclear fire at the center of the Solar System. It is the heat generating engine running the whole show, and without it, all this, everything we do every day, wouldn't be possible. The Sun is a star, like the myriad billion of other stars in the night sky, only this one is our star, our sustenance.

The Sun contains ninety nine percent (99%) of all the mass in the Solar System, therefore it composes most of the matter in the system. Its gravity is strong enough to hold all the planets and interplanetary bodies firmly in their proscribed orbits, from tiny Mercury to the nebulous Oort

Cloud, and perhaps beyond. Scientists have yet to determine just exactly how far out into interstellar space the Sun's influence extends.

The Sun is approximately one hundred and fifty million kilometers (ninety three million miles) from the Earth. It is 1,392,000 kilometers (865,400 miles) in diameter with a volume one million three hundred thousand times greater than the Earth. At its center, the temperature is 15 million degrees Celsius (59 million degrees Fahrenheit) and the pressure is 2.7 billion atmospheres using the equation of hydrostatic equilibrium.

The Sun rotates on its own axis about once every thirty Earth days (one solar day) and it is inclined on its axis to the plane of the ecliptic by seven degrees (7°). The fact the Sun, the generator of this whole system, with more than seven hundred times the mass of all the other bodies in its system put together, is inclined on its axis is intriguingly important, and a blatantly telling indicator of the extent of the disruption in the system in the past. The Agent of Mass Destruction also affected the star at the center of this system, along with every other member of the whole system.

Mercury

Mercury travels along its orbital path at about 58 million kilometers (36 million miles) from the Sun and it makes one revolution in eighty eight Earth days (one Mercurian year). It has a diameter of 4,875 kilometers (3,030 miles) and it rotates on its axis in just under fifty nine Earth days (one Mercurian day). It is tilted on its own axis by two degrees (2°). It is almost vertical in its spin.

Mercury was one of the identified planets from the ancients. The earliest mention of Mercury was in the Assyrian Tablets known as the Mul Apin,

Mercury was named for the god of trade and commerce, but he is better known as the messenger of the gods in Roman Mythology, the equivalent of Hermes in Greek Mythology.

Mercury, the planet, is completely covered by impact craters, and whereas scientists like to assign the impact events to the formative stages of the Solar System, they could just as easily be the result of the proposed ancient celestial opera.

Mercury has no moon.

Venus

Venus orbits the Sun at around 96 million kilometers (about 64 million miles) and makes one revolution in its orbit in two hundred and twenty five Earth days (one Venusian year). It is 40,000 kilometers (about 24,000 miles) in diameter and revolves once on its own axis in about two hundred and forty three Earth days (one Venusian year), very slow in comparison to Mercury, and it rotates retrograde compared to the rotational direction of Earth and all the other planets. Venus rotates clockwise when viewed from its north pole downward, whereas the Earth rotates counter-clockwise. Venus is tilted on its own axis by one hundred seventy seven degrees (177°) (it is nearly upside down in relation to the ecliptic). This may be why it rotates retrograde.

Venus is another planet first observed by our ancient forebears, being the closest planet to the Earth. It was first observed by virtually all early civilizations and due to its close proximity they were even able to observe its transits and phases, like the Moon.

Venus was named for the goddess of love and beauty in Roman Mythology, the equivalent of Aphrodite in Greek Mythology.

Venus, although completely shrouded in dense impenetrable clouds, has recently been pierced by our computer technology, and now reveals tremendous mountains and yawning canyons on the surface of this enigmatic planet. These mountains and canyons could very well be the result of a confrontation with the Agent of Mass Destruction as some of the mountains and canyons on Earth and Mars are evidence of same.

This thesis proposes at some time in the past, Venus occupied a particularly critical place in the Solar System, quite different than at present.

This thesis theoretically proposes the planet we call Venus was once the ravaging and meandering Agent of Mass Destruction, and was exclusively responsible for all the observable damage being meticulously described in this thesis, until it finally and irrevocably settled into its current stable orbit.

This thesis proposes the encounter of Venus (the Agent of Mass Destruction) with all the bodies in the Solar System was monumentally immense, indeed. It literally tore Venus' crust and mantle, literally ripping open the wide abysmal canyons and raising the up-thrusting sky-kissing mountains through the overbearing concentrated cloud layer. It may indeed have twisted the crust on the mantle as there appears to be radiometric evidence of a gaping planet-wide girdling rifted canyon surrounded by vast mammoth mountains. If this is so, then Venus was a planet caught in a vise and this level of rending is evident nowhere else in the Solar System. It speaks of tremendous global geological wrenching of the planet, like one would wring out a sponge. This would be expected if this planet was the Agent of Mass Destruction in the past and it appears this ripping and tearing of Venus happened over a vast period of time, over eons and eons, and repeatedly, again and again. Available evidence points to the fact Venus was the marauder, the interloper, which ended its wandering by finally being captured by the Sun and settled into its orbit, the orbit we see it occupying today.

Venus has no moon.

Earth

Earth revolves around the Sun at 150 million kilometers (93 million miles) and makes one orbital revolution around the Sun in one Earth year (365¼ days). It is 38 thousand kilometers (24 thousand miles) in diameter and rotates on its axis in one Earth day. The Earth is tilted on its own axis by twenty three and a half degrees (23.5°), which causes the four seasons enjoyed by every living thing on Earth each year. It seems the seasons have been running their course since time immemorial, however, the seasons may be a relatively new phenomena, new and refreshing and dazzling.

Earth was named by God. In Roman Mythology its name was Terra, and in Greek Mythology it was named Gaia. Almost all early cultures had a different name for the home planet.

The Earth is replete with convincing evidence of impacts in the past. There is the geologic record, ready at hand to interpret this evidence on the Earth, unlike the other planets and moons where we have to rely on telescopes and satellite images to gather evidence and these methods, though quite satisfactory, are not as defining as first hand investigation.

Scientists know of many of the impact sites on Earth, like at Chicxulub in the Yucatan Peninsula of Mexico, and there is unrelenting evidence in the fossil record of numerous catastrophic collisions in Earth's past, in man's prehistory and even into the historic period as this thesis will show. Scientists estimate the impact at Chicxulub was at sixty five million years ago, and there was another, more devastating extinction event, at two hundred and forty million years ago, called the "Great Dying", with many more impacts of lesser degree in the interim.

There is evidence in the arid deserts of central Arizona of an impact site as early as fifty thousand years ago. In the swampy interior of Argentina, at Rio Cuarto, there is evidence of numerous teardrop shaped impacts at or near the event two hundred and forty million years ago, like the Shoemaker-Levy 9 Comet on Jupiter, when a comet broke up before entry and cascaded down in multiple hits. There is also evidence of this type of impact event, a breakup leaving numerous craters over a wide far reaching area, in the United States of America, from Nevada all the way across the states to Ohio.

There are also impact craters in just about every continent on the planet.

However, there may be other, not as obvious, areas of impact sites, sites which are now covered with water and are perceived as lakes or seas today. This thesis proposes the Hudson Bay in Canada, the Aral Sea in Russia and even the Finger Lakes in upstate New York are all impact sites, and these are just a few of the many sites of this nature.

There is also evidence in North America, from Quebec, Canada to Baja, Mexico, for a major impact site stretching the length of Canada and the United States, from the Ungava Peninsula, which was raped and ruined at some time in the past, to expose and ravish the granite bedrock, so thoroughly it never recovered, no soil ever being deposited there again. Then the scraping Agent went across Hudson Bay, skipping then to the American Badlands, jumping over the Rocky Mountains to plough its way through the Canyonlands of Utah and Colorado, to the Grand Canyon itself and ending at the Baja Peninsula.

The area just described is evidence of another planet sized body actually scraping the Earth along this erratic line, wiping out Lake Agassiz in Ontario allowing the water to escape into the Atlantic Ocean, and draining Lake Bonneville in Nevada into the Gulf of Mexico, and changing many of the prehistoric features on the North American continent.

It slammed the Earth tremendously hard at the end of this scrape, leaving the mile wide gash called the Grand Canyon, a popular tourist attraction today. It may then have kissed the Earth one more time before departing at the Baja Peninsula, forever separating it from mainland Mexico, creating the Gulf of California.

The tremendous gravitational pull of this transforming encounter left its mark on many other areas of the Earth. The table lands of Venezuela were probably dramatically uplifted during this event, as were the Cascade ranges in western North America. On the other side of the planet, there was also evidence of catastrophic activity at this event, the Siberian Traps and the Deccan highlands are just two examples.

There are a number of geologic features in the Yangtze River basin of China which speak to a catastrophic origin, a pushing up of the entire land in this area. The river runs through many high jagged mountain ranges showing evidence of sudden uplift in prehistory.

The whole Tibetan Plateau also shows evidence of a pushing out of this area based on an impact on the other side of the globe. The Tibetan Plateau was upraised at this event very much like the Venezuelan Tablelands and the Ugandan Highlands of Africa.

The origin of the Ural Mountains in Russia may also find its explanation in this event. Scientists conjecture the origin of mountains and mountain chains resulted from plate tectonics. When two plates collide, their leading edges crumple and mountains result. The Ural Mountains are an anomaly in this regard, since no two plates meet at this intracontinental area. However, the Agent of Mass Destruction could have caused the Ural Mountains by catastrophic collision.

The sands of the Sahara also offer mute testimony to the event, desertification in this area, as well as the whole Saudi Arabian peninsula resulting from the scrape on the other side of the globe.

Another effect of the past catastrophic event resulted in the geographic and magnetic north and south poles being displaced from one another. The magnetic field of the Earth has shifted many times in the past, sometimes, like when an event of enormous proportion occurs, the shift is truly immense, as it is currently. It moves year to year based on the magnetic field of the Earth. The magnetic pole can travel up to ten degrees (10°) from the geographic pole.

Whereas the Earth shows much evidence of early catastrophic encounters with another celestial body, the Moon does so even more. The Earth and its enigmatic inscrutable Moon, by their unusual partnership, present anomalies of their own, which are discussed below in the section on the Earth/Moon Enigma.

The Moon

The Moon is 3,476 kilometers (2,160 miles) in diameter and is 386,252 kilometers (240,000 miles) from the Earth. It travels once around the Earth in approximately 27.3 days (one sidereal period). It rotates on its axis in 29.5 days (one synodic period), and because they are both moving around the Sun during these two periods, the Moon always keeps the same side facing the Earth.

Earth's Moon shows much evidence of impact events. The moons of all the planets show impact craters fairly uniformly around their circumference, but the Earth's Moon strays from this universal model. The dark side of the Moon is not as cratered as the side facing the Earth. This is not easily explained considering the side facing the planet is not usually more cratered than the side facing away in fact, it is usually the opposite.

This is evidence of the event described above, the extreme scraping of the North American continent. The Earth and the interloper, at this event were vying for the companionship of the Moon, which this thesis proposes was originally a moon of the other planet. When the interloper came into contact with the Earth, it placed its moon in a position where it was squeezed out of orbit around the other planet, and forced into the debris trail being carried along with the rogue planet. This debris was slammed into the leading edge of the Moon at this time, heavily cratering the side of the Moon which would eventually become the light side as viewed from Earth.

Then the Moon was shoved further along its leading edge where it eventually and quite dramatically synchronized with the Earth's mass and gravitational pull.

At this time, the Earth captured the Moon from its original host and it has resided with us ever since.

The Earth / Moon Enigma

By most accounts, the Moon was born of the Earth. The Earth is larger than the Moon, and this would seem to indicate the Earth, at some time in the distant past, ejected the Moon to its present orbit. Also, the Earth is the host, the Moon is the satellite. This would seem to indicate the Moon was born of the Earth.

There are problems with this hypothesis:

First, the Moon is considerably older than the Earth. The Moon rocks brought back to Earth by the astronauts, were dated at five to six billion years old, whereas the oldest rocks on Earth measure four and a half billion years old. This presents scientists with a major dilemma.

Second, the Moon is composed of some rare silica, found exclusively on only one place on the Earth, on one specific spot on the floor of the Pacific Ocean, and what is more, at only one particular spot on the floor of the Pacific Ocean, near the Marianna Trench, the lowest spot on the ocean floor. If the Earth was the parent of the Moon, it would be more reasonable to assume the rare silica, which makes up much of the Moon, would be found throughout the strata of the Earth.

Third, the Moon has a greater specific mass than the Earth. With this in mind, it would be more correct to state the Earth was born of the Moon, not the other way around, for the body of greater specific mass cannot cleave from the body of less specific mass. However, their sizes also preclude this hypothesis because a smaller body cannot eject a larger one.

And fourth, another factor seemingly contradicting the idea of the Moon as an original member of the Earth/Moon partnership since its formative stages, is because none of the other rocky planets of the inner Solar System have a moon of their own. Mercury has no moon, Venus has no moon and Mars has no moon to speak of, merely two captured asteroids. The inner rocky planets of the Solar System formed no moons when they were coalescing, and would seem to indicate the capture hypothesis for the Earth's Moon is the preferable theory.

So, if the Moon was not directly born of the Earth, where did it come from?

This thesis proposes the Moon was captured by the Earth during this tremendous celestial collision with the Agent of Mass Destruction, when the Moon's original host planet struck the Earth, breaking it in half and then later scraping the North American continent. During this monumental epic collision, the Moon became a satellite of the Earth, being stolen by the Earth from its original host planet (the celestial interloper) during the encounter.

Scientists are loathe to accept the capture theory concerning Earth's Moon because of the relative sizes of both bodies, saying it would take more gravitational power than the Earth possesses to capture a body as large as the Moon. They postulate, in order for this to occur, the two bodies would nearly have to touch and this would create catastrophe beyond anything ever seen, and this is in accord with this thesis. The Earth and the Moon did touch, and the Earth did capture the Moon.

This thesis proposes the reason the rare silica are found in only one place on the floor of the Pacific Ocean, is because it is at this point where the Moon struck the Earth, at a later time, during the Reconstruction, causing it to restart its axial rotation, a rotation which ceased because of the destructive encounter.

Mars

Mars orbits the Sun at a distance of 228 million kilometers (141 million miles) and it confirms one revolution (orbit) around the Sun in about six hundred and eighty seven Earth days (one Martian year). It has a diameter about one sixth the size of Earth, about twice the size of Earth's Moon, 6,800 kilometers (4,200 miles). The surface area is about the same as the land surface area on Earth, however, Mars has no visible ocean, or major source of water. There is much evidence of past water activity on the red planet.

The time it takes Mars to rotate once on its axis is about half an hour longer than Earth's day, twenty four and a half Earth hours (one Martian day). Mars is tilted on its own axis by twenty five degrees (25°).

Therefore Mars also has seasons, though not as pronounced as those on Earth, owing to our atmospheric attributes which are absent on Mars.

The comparison between Earth and Mars, being related in many aspects, is not coincidence or accident. They are explained by the premise offered in this thesis.

Mars is another planet first detected in ancient times by most civilizations. Mars is named for the god of war in Roman Mythology, the equivalent of Ares in Greek Mythology.

Mars also shows evidence of the Agent of Mass Destruction with the largest canyon, Valles Marineris, and the highest volcanic peak, Olympus Mons, in the whole Solar System. These features, colossal as they are, could not easily form by natural Martian geologic processes.

They are more likely a result of an impact with something greater than itself, tearing the Martian surface with tremendous force.

The canyon (Valles Marineris) on Mars is greater than the Great Rift Valley on Earth. It covers an area 700 kilometers (440 miles) wide at places and on Earth would stretch from New York City to Los Angeles, by far larger than anything ever appearing on Earth, and definitely evidence of massive geologic disturbances in the past.

The volcano, Olympus Mons, is the largest in the Solar System, some 26 kilometers (16 miles) high, twice the height of Mount Everest on Earth and covering an area the size of the state of Arizona, and is like a giant pimple on the Martian surface. There are three other volcanoes nearly as large as Olympus Mons running in a line with Olympus Mons, creating a massive bulge on this area of Mars, and there is another volcano in the southern hemisphere 6 kilometers (4 miles) high which has a diameter of over 1,600 kilometers (1,000 miles).

There is also a tremendous impact crater on the southern hemisphere of Mars, larger than any impact crater visible anywhere in the Solar System, measuring 2,000 kilometers (1,250 miles) in diameter.

The Agent causing this impact crater was hundreds of miles across, and must have shook the planet to the very core. This may be why Mars lost its atmosphere and its northern hemispherical crust.

Mars outer crust is distinct of all the planets in the Solar System. At the equator, or nearly so, encircling the entire planet is a massive cliff, thousands of feet in height, separating the southern hemisphere from the northern hemisphere. This cliff is a result of the loss of the outer crust in the northern half due to some catastrophic impact event in the planets past. In the southern hemisphere, the outer crust is intact, but in the north, the outer crust is gone. It is just plain missing. The forces necessary to have caused such a catastrophic smashing of the planet, leaving the crater on the southern hemisphere over 2,400 kilometers (1,250 miles) in diameter, and blowing the outer crust off the whole of the northern half of the planet, were forces not encountered anywhere else in the Solar System, at any time in its history. Only the damage evident on the planet Venus rivals the state of Mars, and Venus, though it encountered almost every celestial body in the Solar System in its role as the Agent of Mass Destruction, is not as badly damaged as the planet Mars. The magnitude of this catastrophic event was something never before encountered, or likely ever to be encountered again, and it is evidence of the Agent of Mass Destruction.

Mars has a thin atmosphere containing many of the elements present in the Earth's atmosphere, carbon dioxide, water vapor, nitrogen and oxygen. At some time in the past, the Martian atmosphere was quite substantial, maybe enough to support life, and from some meteorites found here on Earth (ALH 84001), which originated from Mars, there were possibly some bacterial fossils found. Past life on Mars is definitely probable, and this thesis says so, because Mars was once one and the same with Earth.

Mars also has two distinct polar ice caps probably made of water ice, with maybe some frozen carbon dioxide (dry ice) included. The evidence for water on Mars is unmistakable, and this lends itself to the existence of life on Mars at one time. After the massive collision of the Agent of Mass Destruction with the larger Old Earth, the breakup was not an even one, with equal parts of the Old Earth settling into its component parts. It was instead rather lopsided, with Earth getting all the good things, (air, water, life, etc.) and Mars getting shortchanged, losing its atmosphere early on in the process, which caused it to lose its life, and eventual desertification taking over the whole Martian surface. There may still be water within the crust of the planet.

The two trabants which serve as moons for Mars, Deimos and Phobos, are also testimony of our interplanetary contender, being merely two captured asteroids. Both moons are heavily cratered, evidence of severe bombardment in their history.

Phobos

Phobos has a retrograde orbit taking just a little over one-third of an Earth day (one Phobosian month). It orbits Mars at a distance of just 6,000 kilometers (3,700 miles) from the surface, closer to its host than any other moon in the Solar System. It is small and irregular, just 27 by 19 kilometers (16 by 11 miles) around, and although it is larger than Deimos, it could easily fit inside a medium size crater on Earth's Moon.

Phobos was discovered in *CE 1877* by American astronomer Asaph Hall *CE 1829-1907*.

Phobos was named after the Greek deity, the son of Ares and Aphrodite, the god of horror.

Phobos has a large crater, for its size, indicating an impact nearly shattering the small moon, with cracks evident all over the Phobosian surface from the impact. It is very similar in composition to the asteroids in the outer areas of the Asteroid Belt, nearer to Jupiter, and how it came to orbit Mars is still a puzzle to scientists.

The current thesis explains the enigma, the Agent of Mass Destruction carried it from its original home to the orbit of Mars.

Deimos

Deimos has a more regular orbit around Mars at a distance of 23,500 kilometers (14,100 miles). It takes a little over one and a quarter Earth days to orbit Mars (one Deimosian month). It measures about 15 by 11 kilometers (9 by 6 miles) and is irregular in shape. It is the smallest known moon in the Solar System.

Deimos was named after the Greek deity, the twin brother of Phobos.

Deimos was also discovered in *CE 1877* by Asaph Hall.

Like Phobos, it originated in the outer edges of the Asteroid Belt, near Jupiter.

Jupiter

Jupiter is the largest planet in the Solar System. It is 780 million kilometers (484 million miles) from the Sun and it takes just short of twelve Earth years to make one revolution around the Sun (one Jovian year). Jupiter is approximately 138,000 kilometers (86,000 miles) in diameter and it spins on its own axis in not quite ten Earth hours (one Jovian day). This means the Jovian calendar contains 490,570 days.

Its mass is about two and a half times greater than all the other planets combined. Jupiter is tilted on its own axis by three degrees (3°) which means it is practically vertical in its rotation.

The fact Jupiter, the largest planet in the system with such a large mass, is tilted at all speaks to the tremendous size and speed of the Agent of Mass Destruction. For the Agent to affect Jupiter's axial tilt at all means it was either a very large planet, or it was moving at a tremendous speed through the Solar System. This thesis proposes it was a large planet and it was moving fast as it swept through.

Jupiter was another of the planets known by the ancients. Jupiter was named for the head of all the gods in Roman Mythology, the equivalent of Zeus in Greek Mythology.

Jupiter has a thick gas atmosphere surrounding a rock and iron molten core. There is no solid surface on Jupiter.

There is a large cyclonic storm in the planets southern hemisphere visible from Earth based telescopes called the Great Red Spot measuring over 25,000 kilometers (16,000 miles) across. It has been there for as long as men have been observing Jupiter up close, since the English natural philosopher Robert Hooke *CE 1635–1703* first noticed it in *CE 1664*, over three hundred and fifty years thus far. This is one large hurricane.

Jupiter, with its nine known rings, and sixty-seven known moons showing evidence of cosmic collisions, is mute testimony to the Agent of Mass Destruction.

It was the observation of Jupiter and its four major moons, by the famed Italian astronomer Galileo Galilei's *CE 1564–1642*, visible from his telescope on Earth, which led to the abolition of the prevailing view of his day, a view saying the Earth was the center of the Universe. This dethroning of the prevailing view of the cosmos was known as the Copernican Revolution after famed German mathematician and astronomer Nicholas Copernicus *CE 1473–1543*, because it proved there were other bodies in the Solar System which did not revolve around Earth, but rather exhibited actions independent of Earth's influence, in this case, Jupiter. These were Jupiter's four largest moons and they were forever after known as the Galilean Satellites.

Jupiter has sixty-seven known moons and a series of nine known thin rings. The inner ring has a strongly defined outer boundary at a distance of 128,500 kilometers (8,000 miles) from Jupiter, but the inner boundary is not well defined, with the ring material constantly spiraling down to the Jovian surface. There is another ring just outside the inner ring and extending out to 350,000 kilometers (220,000 miles) from Jupiter. The inner ring revolves counterclockwise (prograde), the normal direction of rotation for planetary companions, but the outer ring revolves clockwise (retrograde).

The outer four moons of Jupiter rotate retrograde, in a direction opposite to the rotation of the host planet, another anomaly which is not easy to explain, unless they were captured by Jupiter, from the calamitous Agent, as it swept by Jupiter in a retrograde direction. There are many moons, in our Solar System, which rotate retrograde and they can all be explained by this ruinous incident.

The Four Inner Moons

METIS - Metis travels in orbit just 127,960 kilometers (79,510 miles) from Jupiter which places it between the inner two rings. It is one of the Shepherd moons, keeping the rings intact and circling. It is only about 35 kilometers (20 miles) wide and is very irregular in shape. It is most probably a captured asteroid.

ADRASTEA - Adrastea is the second moon out from Jupiter at 128,980 kilometers (80,140 miles) which makes it another shepherd moon orbiting between the two rings.

Adrastea is also shaped very irregularly and is only slightly larger than Metis. It also is probably a captured asteroid.

AMALTHEA - Amalthea orbits Jupiter at 181,300 kilometers (112,700 miles) or within the first two rings gap. Amalthea is the largest of the inner moons (the shepherd moons) at 135 kilometers (84 miles) in diameter. It, too, is probably a captured asteroid because it also has an irregular shape, not spheroid, and is quite small.

THEBE - Thebe orbits Jupiter at 221,900 kilometers (137,900 miles) from the Jovian surface, again within the first gap between the inner rings. It is also an irregularly shaped chunk of rock smaller than Metis, and like the other four shepherd moons, is probably a captured asteroid.

The Four Major Galilean Satellites

The four moons first observed by Galileo are named in his honor and are known as the Galilean moons. They are the next four moons in order out from the planet and are the four largest of the Jovian satellites, ranging in size from as large as the Earth's Moon to as large as the planet Mercury.

IO - Io is in orbit around Jupiter at 421,000 kilometers (262,000 miles) and takes just one and three quarters Earth days to make one revolution around its host planet (one Ionian month). It is 3,630 kilometers (2,260 miles) in diameter, just slightly larger than Earth's Moon. It was named for one of the mythical lovers of Zeus in Greek Mythology, whom the Romans called Jupiter. It is composed of rock and mineral like the inner planets and it is the most volcanically active body in the entire Solar System.

The tremendous volcanic activity of Io is possible evidence of a close encounter with the Agent of Mass

Destruction. In the past, the Agent swept close to Io and stirred up an active molten core causing lava to spew to the surface and it has been flowing ever since, although its close proximity to Jupiter may also be sustaining the rampant volcanism on Io. Io is in a particularly inconvenient place in the Jovian system, constantly being pulled and tugged by Jupiter's immense gravitational pull as well as the tug from Europa and Ganymede, constantly putting internal pressure on the already molten core. This is why Io is the most active geologically of all the bodies (moons and planets) of the Solar System.

EUROPA - Europa is in orbit around Jupiter at 671,000 kilometers (417,000 miles) and it completes one orbit around Jupiter in a little over three and a half Earth days (one Europan month). It is 3,140 kilometers (1,880 miles) in diameter just slightly smaller than Earth's Moon. It is composed of silicate rock and mineral, like the inner planets, and is surrounded by water ice. Below this ice could be an ocean of water or slushy ice. It was named for one of the mythical lovers of Zeus in Greek Mythology. Zeus and Europa were the parents of Minos.

GANYMEDE - Ganymede is in orbit around Jupiter at 1.07 million kilometers (660,000 miles) and takes just over seven Earth days to make one revolution around its host planet (one Ganymedan month). It is 5,260 kilometers (3,160 miles) in diameter, the largest moon in the Solar System, larger than the planet Mercury. It is composed of low density material, similar to the outer gas giants, made mostly of frozen water. It was named for one of Zeus's lovers in Greek Mythology. Ganymede's solid ice crust is 75 kilometers (45 miles) thick.

CALLISTO - Callisto is in orbit around Jupiter at 1.88 million kilometers (1.11 million miles) and takes slightly over sixteen and a half Earth days to make one revolution around its host planet (one Callistoan month). It is 4,800 kilometers (3,000 miles) in diameter, nearly the size of Mercury. It is composed of low density water ice, like Ganymede. It was named for one of the mythical lovers of Zeus in Greek Mythology.

The Outer Moons

There are two families of outer moons orbiting Jupiter at great distances. The first family (the Himalia Group) orbits at a mean distance of 11 million kilometers (6.6 million miles) from Jupiter. The second family is composed of three groups (the Ananke Group, the Carme Group and the Pasiphae Group) and it orbits at a mean distance of 21 to 23 million kilometers (13 to 14 million miles) from Jupiter.

The First Family of Outer Moons

These are known as the Himalia Group and orbit Jupiter at a mean distance of 11 million kilometers (6.6 million miles) from their host planet.

LEDA - Leda is the smallest satellite of Jupiter. It is just 16 kilometers (9 miles) in diameter. It period of rotation is unknown revolving around Jupiter in about two hundred and thirty nine Earth days (one Ledan month). It was named after a lover of Zeus, the Mother of Castor and Pollux, Clytemnestra and Helen of Troy. It is most probably another of the captured asteroids.

HIMALIA - Himalia is 186 kilometers (131 miles) in diameter and like the other eight outer satellites, there is very little else known about them. Its periods of rotation is about twelve Earth hours (one Himalian day) and its period of revolution is two hundred and fofty Earth days (one

Himalian month). Himalia is named after a nymph who bore three sons to Zeus. It also is most probably another captured asteroid.

LYSITHEA - Lysithea has a diameter of 36 kilometers (20 miles). Its rate of rotation is unknown and its orbital revolution is two hundred and fifty nine Earth days (one Lysithean month). Lysithea was named after the daughter of Oceanus and she was one of Zeus' lovers. Obviously another of the captured asteroids.

ELARA - Elara has a diameter of 76 kilometers (45 miles) and it rotates on its own axis in about twelve Earth hours (one Elaran day). It takes about two hundred and sixty Earth days to make one orbit around Jupiter (one Elaran month). Elara was also named after one of the lovers of Zeus, who bore him a son, Tityus the giant. This moon is just another of the many captured asteroids Jupiter has imprisoned over the eons.

The Second Family

These moons rotate retrograde around Jupiter and they are known as the Ananke Group, the Carme Group and the Pasiphae Group (Sinope is usually included in the Pasiphae Group). They travel at a mean distance from Jupiter of 22 million kilometers (13.5 million miles).

ANANKE - Ananke is 30 kilometers (18 miles) in diameter. Its period of rotation is unknown. It takes just over one and a half Earth years to complete one orbit (one Anankean month). Ananke was named after another of the many lovers of Zeus, she was the mother of Adrastea. It is a captured asteroid, no doubt.

CARME - Carme is about 40 kilometers (22 miles) across. Its period of rotation is unknown and it takes almost two Earth years to complete an orbit around Jupiter (one

Carmean month). Carme was also named after one of Zeus' lovers, the mother of Britomartis, a Cretan goddess. It is assuredly another captured asteroid.

PASIPHAE - Pasiphae is about 50 kilometers (32 miles) wide and its period of rotation is unknown at present. It makes one orbit around Jupiter in two Earth years (one Pasiphaen month). Pasiphae was named after the daughter of Helios, the sun god, and she was the wife of Minos, and the mother of Ariadne. It also is just a captured asteroid.

SINOPE - Sinope is about the same size as Lysithea of the first family, 36 kilometers (18 miles) in diameter. Its rotational period is not known at present. It makes one orbit around Jupiter in just over two Earth years (one Sinopean month). Sinope was named for an ancient city which is now situated in present day Turkey. It is the last of the captured asteroids labeled as moons of Jupiter.

This was not an all-inclusive list of Jupiter's moons but it gives a fair sampling of all of them. The largest moon in the Solar System is among these and most of Jupiter's sixty-seven moons are mere asteroids.

Saturn

Saturn is 1.42 billion kilometers (888 million miles) from the Sun and it completes one orbit in not quite twenty nine and a half Earth years. (one Saturnine year) It is 119,300 kilometers (74,130 miles) in diameter and completes one revolution in slightly more than ten and a half Earth hours. (one Saturnine day) Saturn is a massive ringed planet, the second largest in the Solar System.

Its density is less than water. If a large enough body of water existed to plunge Saturn into, Saturn would float. Saturn is tilted on its own axis by twenty seven degrees

(27°), a little bit more tilted than the Earth or Mars, and a tremendous tilt for a planet of such size.

Saturn was another planet known by the ancients and named for the god of agriculture in Roman Mythology, the father of most of the other major gods, the equivalent of Cronos in Greek Mythology.

Saturn, a planet best known for its magnificent ring system, has always shown man evidence of a possible early collision in its illustrious history. This evidence is the rings, one of the most glorious and beautiful features of any of the planets of this Solar System. The evidence has been there ever since man first viewed Saturn through a telescope.

At first Saturn was thought to have seven rings, but further observation has revealed there are more than 100,000 ringlets, and each ringlet circles the planet in its own orbit. The seven rings were named with letters of the alphabet and they are from the planet outward, D C B A F G and E, alphabetized in order of their discovery.

Saturn's Moons

Saturn has sixty-two known moons with confirmed orbits. Many of Saturn's moons show evidence of catastrophic encounters during their illustrious histories, and although many could have been encounters with Saturn herself, others could be proof of the Agent of Mass Destruction.

The Six Prominent Inner Moons

MIMAS - Mimas is about 186,000 kilometers (116,000 miles) from Saturn and it orbits Saturn in twenty three Earth hours (one Mimasian month). It is 392 kilometers (235 miles) in diameter, about one ninth the size of Earth's

Moon. It is spherical in shape, composed mostly of water ice. Mimas is heavily cratered, evidence of bombardment from space, and it has one crater of enormous size, 132 kilometers (78 miles) wide, covering approximately one third of the moon's surface. The impact which caused this crater must have nearly shattered this small moon, very much like Miranda, Uranus' moon. Mimas was named for a Titan slain by Hercules in Greek Mythology. At 390 kilometers (235 miles) in diameter it would qualify as one of the large asteroids, captured by Saturn, but for its spheroid shape. This shape tends to lend credence to Mimas as a moon in its own right.

ENCELADUS - Enceladus is 238,000 kilometers (143,000 miles) from Saturn and it orbits Saturn in about one and a half Earth days (one Enceladean month). It is 500 kilometers (310 miles) in diameter, about one sixth the diameter of Earth's Moon. It is the sixth largest moon of Saturn. It is spherical in shape, composed mostly of water ice with a small rocky core. Enceladus was named for a Titan defeated by Athena in Greek Mythology.

Although Enceladus is heavily cratered, it is less so than the other four inner moons, indicating either it is younger than the others, or it was not in a position to be affected by a pass from the Agent of Mass Destruction in past ages.

However, Enceladus is tectonically active and this may be why it has such a young surface and also may be possible evidence of celestial encounters. Tectonic activity on any of the moons in the Solar System may be evidence of past celestial encounters with the Agent, the clashes of the planet sized body with the moon creating activity in the cores of the smaller moons and stirring up tectonic processes. This may also be true for the planets themselves.

In *CE 2004*, the Cassini satellite did a close flyby, within 1,500 kilometers (932 miles) of Enceladus and found geysers spewing liquid water from the South Polar Region extending over six miles into the thin, wispy atmosphere. This caused a reassessment of the mission and another flyby, in *CE 2008*, within 48 kilometers (30 miles) of the surface, through the geysers. This second flyby detected a salinity of the water, comparable to Earth's oceans, and possibly simple hydrocarbons, which may indicate an extremely good chance there is life in some form on this small moon.

TETHYS - Tethys orbits Saturn at 295,000 kilometers (177,000 miles) in just under two Earth days (one Tethysian month). It is 1,060 kilometers (636 miles) in diameter which makes it about one third the diameter of Earth's Moon. It is spherical in shape, composed mostly of water ice with a small rocky core. It was named for a Titan in Greek Mythology.

Tethys is heavily cratered, and it also has a large impact crater called Odysseus which is 400 kilometers (240 miles) wide, covering two fifths of Tethys surface. Scientists theorize the impact making this crater should have completely shattered Tethys, and scientists conclude the impact must have occurred when Tethys was still in a liquid state. This theory cannot be true for obvious reasons. If Tethys was liquid at the time of the impact, the crater would never have formed because of the nature of liquidity. This is like saying the surface of a lake will be imprinted with the impact of a rock thrown into it. Whereas this is true and the imprint of the impact of the rock on a lake is visible, it does not remain so, not even if the lake freezes after the impact, since freezing is a gradual process, unless it was flash frozen instantaneously. If this was so, what

Agent could have accounted for this flash freezing of an entire moon?

It also has a very large valley 100 kilometers (62 miles) wide and over 2,000 kilometers (1,200 miles) long. A surface feature of this proportion is evidence of major geologic activity in the past, possibly a result of the Agent of Mass Destruction. It appears Tethys, like Miranda and many of the other moons of the outer Solar System, were visited in the past by the Agent of Mass Destruction and the surface features are a result of this visitation. Tethys was nearly destroyed by this encounter and the moon as it stands today is stark evidence of our smashing intruder.

TELESTO AND CALYPSO - These two tiny moons of Saturn also orbit in the orbit of Tethys. They are therefore called the Tethys Trojans. Telesto leads Tethys about sixty degrees (60°) in the orbit and Calypso, about sixty degrees (60°) behind. Both moons are irregularly shaped and both revolve at the same speed as Tethys, about two Earth days (one Telestoan and one Calypsoan month). Both are about 30 kilometers (20 miles) in diameter, mere lumps of rocky material. They are not round and therefore qualify as mere trabants, captured asteroids. Due to their sizes and positions, and the shattered nature of Tethys, these two small moons may once have been part of Tethys.

DIONE - Dione is 377,000 kilometers (234,000 miles) from Saturn and makes its orbit in three Earth days (one Dionean month). It is 1,120 kilometers (672 miles) in diameter, about one third as large as Earth's Moon. It is spherical in shape, composed of one third rock and two thirds water ice. Dione is named for the mother of Aphrodite in Greek Mythology. It is not an asteroid, owing to its spheroid shape. Dione's dark side, the side facing away from Saturn, is heavily cratered, evidence of bombardment from space. On most moons the side facing away from the planet is the side most heavily cratered, as is Dione, and unlike the Earth's Moon. This is more evidence of the Agent. It has two large craters on its surface, one named Aeneas is 150 kilometers (93 miles) across, and the other named Dido is 125 kilometers (78 miles) across.

HELENE - Helene is the same distance from Saturn as Dione, since both share the same orbit around their host planet. Helene is Dione's Trojan moon. It is about 32 kilometers (20 miles) in diameter and though its rotational speed is unknown, it revolves around Saturn in the same orbit as Dione, with the same revolution, about three Earth days (one Helenean month). Little else is known about this moon at the present time. The launch and explorations of the Cassini probe will hopefully reveal more about these lesser known moons.

RHEA - Rhea is the largest airless satellite of Saturn. Rhea orbits Saturn at 527,000 kilometers (327,000 miles) and makes one revolution in four and a half Earth days (one Rhean month). It has a diameter of 1,530 kilometers (931 miles) which makes it about half the size of the Earth's Moon. Rhea is Saturn's second largest moon and is spherical in shape, composed mostly of water ice with a rock core making up approximately one third of the moon, very much like Dione. Rhea was named after the sister and wife of Cronos in Greek Mythology. Rhea is very like the moon Dione, although larger. It too is heavily cratered on its dark side, and there is a lot of evidence to indicate Rhea and Dione may once have composed a single body. There are numerous impact craters on Rhea's dark side, some as wide as 40 kilometers, (25 miles) which tell of a massive bombardment from space. The two largest craters are 100 kilometers (80 miles) and 80 kilometers (50 miles) wide, respectively.

Other Inner Moons

These moons are located within the rings system of Saturn and they act as shepherds of the rings, keeping them intact and rotating around Saturn. They are all irregularly shaped and probably began life as asteroids before their promotion to moons by the gravitation pull of Saturn. Not much is known about them at present.

ATLAS - Atlas circles Saturn just outside the A ring. It is only 40 by 20 kilometers (25 by 15 miles) in circumference. Its rate of rotation is not known but its orbital period is just over thirteen Earth hours (one Atlasian month). It was named for the son of Iapetus, the brother of Prometheus.

EPIMETHEUS - Epimetheus is 144 by 108 by 98 kilometers (89 by 67 by 61 miles) in circumference, an irregularly shaped moon orbiting its host in just fourteen Earth hours (one Epimetheusian month), the same amount of time it takes to make one rotation on its axis (one Epimetheusian day). This means it always keeps the same side to its host planet, Saturn, like Earth's Moon always keeps the same side facing the Earth. The surface is heavily cratered with some very large craters being nearly 30 kilometers (20 miles) in diameter. Its surface is also scarred by several ridges and valleys which attest to frictional forces operating on the moon, more than can be accounted for by the ring it shepherds or the planet it orbits. Epimetheus was named after the son of Iapetus, the brother of Prometheus and Atlas. Epimetheus and Janus share the same orbit around the host, being only about 50 kilometers (35 miles) apart. About once every four Earth years, Janus and Epimetheus approach each other in their orbit and change places, one becoming the inner moon and the other becoming the outer moon, and vice versa. This may be an indication at some time in the past, these two small moons were once a larger body, and some outside influence (the Agent of Mass Destruction) broke it apart to form this mutual orbit of two separate moons.

JANUS - Janus is irregular in shape with a diameter of about 190 kilometers. (120 miles) It revolves around Saturn in the same time and orbit as Epimetheus, and like Epimetheus, always keeps the same side to Saturn, which means its rate of axial rotation is the same as the rate of its orbit (one Janusian month and one Janusian day). Janus shares its orbit with Epimetheus as described above. It is heavily cratered with some large craters measuring 30 kilometers (20 miles) in diameter, but unlike Epimetheus, it has no surface tearing scars. Janus was named after the god of gates and doorways, the god with two faces always looking in opposite directions.

PAN - Pan is the innermost known satellite of Saturn, revolving in the Encke Gap of the A ring, keeping it open. It is only about 20 kilometers (15 miles) in diameter and moves once through its orbit around Saturn in just over twelve Earth hours (one Panian month). Pan was named for the god of shepherds, an appropriate name for the leader of the shepherd moon.

PANDORA - Pandora and Prometheus revolve in nearly the same orbit around Saturn, near the F ring. It takes Pandora about fourteen Earth hours to make one revolution around Saturn (one Pandoran month). Pandora is the shepherd for the outer edge of the ring and Prometheus for the inner edge. Pandora is shaped very irregularly also, being 114 by 84 by 62 kilometers (71 by 52 by 38 miles) in circumference and if it rotates on its own axis it is unknown at this time. The surface of Pandora is heavily cratered, owing to the place it holds in the Saturnine system, being bombarded by objects from the rings continuously during its life. It has two large craters about 30 kilometers (19 miles) in diameter. Pandora was named for the first woman on Earth in Greek Mythology, the woman who opened the forbidden jar (not a box) and released upon man all forms of disease, pestilence and evils.

PROMETHEUS - Prometheus acts as shepherd for the inner edge of Saturn's F ring. It is extremely elongated and very irregular in shape, being about 145 by 85 by 62 kilometers (90 by 53 by 39 miles) in circumference. Its period of rotation is unknown at present but it revolves in an orbit in just over thirteen Earth hours (one Prometheusian month). It is cratered with some larger craters measuring 20 kilometers (12 miles) in diameter. This is possible evidence of the close encounter with the Agent. Given the place in the ring system of Saturn, it is more probably explained by the ring. Prometheus is probably made up of water ice. Prometheus was named for the Titan who was the benefactor of humanity, giving fire to man. He was chained to a rock for this offense, where an eagle was assigned to peck at him, until Hercules rescued him.

S 1995 S3 - This newly discovered moon is one of the shepherd moons lying just outside the F ring. It is only about 25 kilometers (15 miles) across. It lies very close to the moon Epimetheus. It has not yet received a name.

S 1995 S4 - This moon was discovered at the same time as S3. It lies about 10,000 kilometers (6,000 miles) further out from the F ring than S3. It has also not yet received a name.

Trojan Satellites

Also there are possibly four Trojan satellites of Tethys and one of Dione thus far discovered, though not yet confirmed. Trojan satellites are those sharing the same orbit as the main moon, either leading the moon or trailing it in the orbital path.

Titan

Titan is Saturn's largest moon, and the second largest moon in the Solar System after Jupiter's moon Ganymede. It orbits between the inner moons and the outer moons. It is 5,150 kilometers (3,200 miles) in diameter, larger even than the planet Mercury. It orbits at a distance of 1.22 million kilometers (750,000 miles) from Saturn and it completes its orbit in sixteen Earth days (one Titanian month). It is composed of a rocky core approximately 3,400 kilometers (2,100 miles) in diameter surrounded by water ice. The interior is probably molten. Titan was named for the family of giants in Greek Mythology known as the Titans.

Although Titan is covered by a visibly impenetrable orange cloud layer, scientists were able to make out some surface features through heat refraction measurements and they have identified an area of elevation roughly the size of Australia on the surface. This area could be a continent in Titan's vast ocean of methane. The Cassini orbiter will send a probe, the Huygens probe, into Titan's atmosphere in *CE 2004* and more evidence about this enigmatic moon should then be available.

The Outer Moons

HYPERION - Hyperion orbits Saturn at a distance of 1.48 million kilometers (888,000 miles) and completes one orbit in twenty one Earth days (one Hyperionian month). It is irregular in shape, being 410 kilometers (250 miles) at its widest point and 220 kilometers (130 miles) at its narrowest point. It is composed mainly of water ice. It is named for one of the Titans in Greek Mythology. It is heavily cratered with some craters measuring up to 120 kilometers (75 miles) wide and 10 kilometers (6 miles) deep, nearly one third of its diameter. Planetary bodies this size and larger, as a rule, are usually impelled by gravity and rotation to be spheroid, a sphere being the most compact shape. Hyperion defies this rule, very much like Neptune's moon Proteus. Scientists hypothesized Hyperion was once spheroid but was changed to an irregular shape by a collision, just as proposed in this thesis. They say the collision was with an asteroid or a comet whereas the thesis says it was with the Agent of Mass Destruction.

IAPETUS - Iapetus moves in an orbit 3.56 million kilometers (2.21 million miles) from Saturn, and it completes its revolution in twenty one Earth days (one Iapetusian month). It is 1,460 kilometers (905 miles) in diameter which is the third largest of Saturn's moons, only about a third the size of Earth's Moon. It is composed almost entirely of water ice with a very bright surface interrupted by a large dark area. Iapetus was named after one of the Titans, a son of Uranus and Gaea, in Greek Mythology. The dark area is on the side of the moon facing the direction of its orbit. The other side is exceedingly bright. There are four major craters each measuring about 100 kilometers (60 miles) across, showing evidence of the same sort of collisions many of the other moons of the outer Solar System show.

PHOEBE - Phoebe is the only moon of Saturn to travel in a retrograde orbit around Saturn at a distance of 13 million kilometers (8 million miles), slightly more than four times farther out than Iapetus, the next closest moon to Saturn. Phoebe is the farthest of Saturn's large moons from its host. Phoebe takes five hundred and fifty Earth days to complete one revolution around its host planet (one Phoebean month). It has a diameter of 220 kilometers (132 miles) which is only one fifteenth the size of Earth's Moon. It is named after a goddess in Greek Mythology who was the goddess of the hunt and of the moon. By its density, it can be determined it is either composed mostly of water or it is filled with large cavities. It is probable Phoebe was a captured comet, maybe one of the Centaurs, a group of comets or asteroids in orbit around the Sun between Saturn and Uranus.

Uranus

Uranus has a diameter of 51,120 kilometers (31,771 miles) and it is 2.87 billion kilometers (1.78 billion miles) from the Sun. Uranus takes eighty four Earth years to complete one rotation around the Sun (one Uranusian year). It rotates on its own axis in just over seventeen Earth hours (one Uranusian day). Uranus is inclined on its own axis to the plane of the ecliptic by ninety eight degrees (98°), or in other words, it is lying on its side. Uranus, at some time in the distant past, and by some unknown agent was knocked completely on its side, the only planet in the Solar System oriented this way.

The size of this planet is immense and the force necessary to tip it over was monumental indeed, similar to the force needed to knock the outer crust off Mars' northern hemisphere, indicating either a large protagonist, or one travelling at a tremendous speed, or perhaps both factors were involved. The fact Uranus rotates on its side is compelling evidence for the Agent of Mass Destruction.

Uranus was discovered in *CE 1781* by astronomer William Herschel *CE 1738-1822*. Uranus was named for the god of the sky in Greek Mythology, the equivalent of Erebus (Chaos) in Roman Mythology.

Uranus has nine known rings and twenty-seven known moons, most named for characters in the works of William Shakespeare *CE 1564-1616 and* Alexander Pope *CE 1688-1744.*

Uranus' nine rings are further evidence of the celestial marauder. The rings were captured by Uranus, as all rings were captured by their host planets, from the debris left by the destructive sojourner.

Uranus' moons present more intriguing evidence, many of them show evidence of close contact with the detrimental interloper.

Two of Uranus' twenty-seven moons revolve in retrograded orbits, which as previously shown was perhaps the result of the Agent of Mass Destruction. And in Miranda, there is evidence of a planetary member which inadvertently strayed directly into the path of this juggernaut. Scientists, upon viewing Miranda for the first time, described it as a moon which looks like it was shattered and reassembled itself, and not too carefully, appearing fragile at best in its reconstructed state.

Uranus' Moons

The five major moons of Uranus are large enough to have achieved hydrostatic equilibrium (q.v.).

TITANIA - Titania is the largest of Uranus' moons measuring 1m580 kilometers (947 miles) about half the diameter of Earth's Moon. It orbits Uranus at 436,000 kilometers (262,000 miles), and it completes its orbit around Uranus in a little less than nine Earth days (one Titanian month). Titania was discovered by Herschel, along with Oberon in *CE 1787.* Its surface is about half ice and half rock, with light areas on the surface. There is little evidence of cratering, meaning either it has a relatively young surface, or it was situated in a position around Uranus where it incurred little damage during the Agent of Mass Destruction's transit through this portion of space in the distant past. There is much evidence the ice on Titania has melted and refrozen over the centuries, and perhaps this melting is the result of friction caused by limited contact with the Agent. Titania is named for the queen of the fairies in Shakespeare's *"A Midsummer Night's Dream"*.

OBERON - Oberon is the second largest of Uranus' moons measuring 1,520 kilometers (940 miles) again about half the diameter of Earth's Moon. It orbits Uranus at 583,000 kilometers (350,000 miles) making it the farthest moon away from the host planet, and it completes its orbit around Uranus in just thirteen and a half Earth days (one Oberonian month). It also was discovered by Herschel. Its surface is black and heavily cratered showing evidence of past buffeting from space. There are lighter areas of ice underneath which show through where the ice has cracks, more evidence of geologic activity on this moon. Oberon is named for the king of the fairies in Shakespeare's *"A Midsummer Night's Dream"*.

UMBRIEL - Umbriel is the third largest of Uranus' moons measuring 1,170 kilometers (702 miles) about one third the diameter of Earth's Moon. It orbits Uranus at 266,000 kilometers (160,000 miles), and it completes its orbit around Uranus in just four Earth days (one Umbrielian month). Umbriel and Ariel were both discovered in *CE 1851* by English merchant and amateur astronomer William Lassell *CE 1799-1880*. Umbriel is half ice and half rock, like Titania, but it has the darkest surface of any of Uranus' moons, and its surface is heavily cratered. There is one interesting area where a surface feature resembles a light colored ring over 140 kilometers (84 miles) across, named Wunda. This ring on the surface may be a result of a collision with the Agent, as the heavy cratering most certainly could be also. Umbriel was not named after a Shakespearean character. It was named after a character in the poem *"The Rape of the Lock"* by the CE 18th century British poet, Alexander Pope.

ARIEL - Ariel is the fourth largest of Uranus' moons measuring 1,160 kilometers (721 miles) about one third the diameter of Earth's Moon. It orbits Uranus at 185,000 kilometers (116,000 miles), and it completes its orbit around Uranus in just two and a half Earth days (one Arielian month). Ariel, too, was discovered by Lassell. Ariel is half ice and half rock, like Titania and Umbriel, and it has the lightest surface of any of Uranus' moons. Its surface has many, many old craters and the ice is severely cracked, with cracks up to hundreds of kilometers long and tens of kilometers deep. The face of Ariel is severely scarred, showing evidence of much turmoil on its surface over the millennia. Ariel is named after a spirit who saves Prospero from Sycorax in Shakespeare's *"The Tempest"*.

MIRANDA - Miranda is one of the strangest moons of Uranus measuring 480 kilometers (298 miles) about one eighth the diameter of Earth's Moon. It orbits Uranus at 130,000 kilometers (77,900 miles), and it completes its orbit around Uranus in just thirty four Earth hours (one Mirandan month). Miranda was discovered in *CE 19*48 by Gerard Kuiper, who also discovered the Kuiper Belt. It also is half ice and half rock, like Titania, Umbriel and Ariel, and it has the most interesting surface of any of Uranus' moons. The face of Miranda is severely scarred, showing evidence of the Agent of Mass Destruction, and destruction is the key word on Miranda. When scientist first viewed the Voyager 2 photographs sent back of Miranda, they described the moon as being blasted apart and reformed numerous times. This terminology does not imply an internal mechanism, but rather an external one. The Chevron and the Racecourse are just two massive areas, among many, where the surface of Miranda was ripped and torn by extreme geologic twisting. Miranda also has cliffs over 15 kilometers (9 miles) high and canyons 20 kilometers (12 miles) deep. These cliffs and canyons are evidence of severe outside tensions being exerted on the tiny moon Miranda. Miranda is named after the daughter of Prospero in Shakespeare's *"The Tempest"*.

CORDELIA, OPHELIA, BIANCA, CRESSIDA, DESDEMONA, JULIET, PORTIA, ROSALIND, BELINDA AND PUCK

These ten moons of Uranus are small and not very well researched because they were only discovered by the Voyager 2 flyby and they were not photographed well during any of man's ventures to the outer planets.

Cordelia has a diameter of 40 kilometers (25 miles and was named after the tragic heroine in Shakespeare's *"King Lear"*.

Ophelia is 42 kilometers (26 miles) in diameter and was named for the daughter of Polonius in Shakespeare's *"Hamlet"*.

Bianca measures 54 kilometers (34 miles) in diameter and was named for Cassio's jealous lover in Shakespeare's *"Othello"*.

Cressida has a diameter of 82 kilometers (51 miles) and was named after the heroine in Shakespeare's *"Troilus and Cressida"*.

Desdemona measures 70 kilometers (43 miles) in diameter and was named after the Venetian beauty in Shakespeare's *"Othello"*.

Juliet is 106 kilometers (66 miles) in diameter and was named for Juliet, the daughter of the merchant Capulet in Shakespeare's *"Romeo and Juliet"*.

Portia has a diameter of 140 kilometers (87 miles) and was named for the heroine in Shakespeare's *"The Merchant of Venice"*.

Rosalind measures 72 kilometers (45 miles) in diameter and was named for the heroine and protagonist in Shakespeare's *"As You Like It"*.

Belinda measures 90 kilometers (56 miles) in diameter and was named for the owner of the lock in Alexander Pope's *"The Rape of the Lock"*.

Puck has a diameter of 162 kilometers (101 miles) and was named for the sprite Robin Goodfellow who was called Puck in Shakespeare's *"A Midsummer Night's Dream"*.

The photographs sent back in *CE 1986* show much the same evidence of collisions as the five large moons show. Uranus and its moons were definitely visited by the Agent of Mass Destruction at some time in the distant past, perhaps many times.

PROSPERO S/1999 U3, SETEBOS S/1999 U1 AND STEPHANO S/1999 U2

These three were discovered in *CE 1999* from the photograph sent back by Voyager 2 in *CE 1986*.

Prospero is 50 kilometers (31 miles), Setebos is 48 kilometers (30 miles) and Stephano is 32 kilometers (20 miles) in diameters respectively, and this is pretty much all that is known of these three moons. They were sufficiently large to be named immediately but they are small and it is assumed they are merely captured asteroids.

Prospero is named for the rightful Duke of Milan in Shakespeare's *"The Tempest"*. Setebos was named for the deity worshipped by the witch Sycorax in Shakespeare's

"The Tempest". Stephano was named for the boisterous drunken butler in Shakespeare's *"The Tempest"*.

CALIBAN AND SYCORAX (S/1997 U1 and S/1997 U2)

Scientists once thought there were no irregular moons orbiting Uranus, until scientists at the Hale Observatory on Mount Palomar discovered two in *CE 1997*. Caliban and Sycorax were discovered by a team headed by Brett J. Gladman born *CE 1966*. Caliban is 72 kilometers (45 miles) in diameter and Sycorax is 150 kilometers (93 miles) in diameter.

They named these two Caliban and Sycorax, Caliban was named after a monster, and Sycorax after an unseen witch, both in *"The Tempest"* by Shakespeare.

Little is known about these two irregular shaped moons, other than they orbit far outside the other moons and are in angular orbits. They are also most probably captured asteroids.

PERDITA – (1986 U10)

This satellite was also found in *CE 1999* from the Voyager 2 photographs of *CE 1986*. It was located on the photograph by Erich Karkoschka born *CE 1955*. Perdita is 26 kilometers (16 miles) in diameter. It was named Perdita, which is Latin for "lost" and was a character, the daughter of Leonties and Hermoine in *"The Winter's Tale"* by Shakespeare.

It is merely a rock, bringing the total of Uranus' satellites to twenty-seven thus far discovered.

Neptune

Neptune is the eighth planet out from the Sun. It is 4.5 billion kilometers (2.8 billion miles) from the Sun and it completes an orbit around the Sun in just short of one humdred and sixty five Earth years (one Neptunian year). It is 49,400 kilometers (30,700 miles) in diameter and it completes one rotation in just sixteen Earth hours (one Neptunian day). It is tilted on its own axis by almost thirty degrees (30°), slightly more tilted than Saturn, another planet affected by the mischievous marauder.

Neptune was discovered in *CE 1846* by the French mathematician Urbain Le Verrier *CE 1811-1877*. Neptune was named for the god of the sea in Roman Mythology, the equivalent of Poseidon in Greek Mythology.

Neptune, the planet, shows no other obvious evidence of interaction with this ravaging celestial spoiler, however Neptune has a gaseous atmosphere, as do Jupiter, Saturn and Uranus, and it is difficult to see any evidence on any of these gas giants of a catastrophic past beneath this dense cloud layer. Neptune, like Jupiter, has vast storms in the clouds on the surface, one named the Great White Spot as opposed to Jupiter's Great Red Spot. Both are large and maybe perpetual hurricanes of fantastic strength. Neptune has winds in its atmosphere travelling at thousands of kilometers per hour, ferocious winds that constantly swirl around the whole planet.

Neptune has five known rings and fourteen known moons.

Though Neptune shows little evidence of the interloper, the Agent of Mass Destruction, the moons of Neptune give evidence of catastrophe, as do the moons of all the gaseous planets, a lingering result of this Agent of Mass Destruction.

The Rings

Neptune's five known rings are probably the remains of a tremendous cosmic catastrophe. This thesis proposes all the rings in the Solar System, those of

Neptune, Uranus, Saturn and Jupiter, are the remains of crushed and shattered larger bodies captured by the host planets over the geologic centuries and pulverized to their present sizes by the actions of their host planet and the other orbiting debris in their vicinity.. Some rings are massive as is the case with Saturn, and some not so, as with the other three gas giants. All the gas giants have rings.

The rings of Neptune may be the remains of a larger body impacted by the Agent of Mass Destruction and it is entirely possible Pluto and Charon are also remnants of this larger body. Pluto and Charon being rather large pieces of a once large moon of Neptune which got shattered in an early encounter with the Agent of Mass Destruction and flung away from Neptune's gravitational influence to its present orbit. The rings are the residual components of this collision which were captured by Neptune over time.

Neptune's Moons

Neptune has fourteen known moons.
TRITON - Triton is the largest of the moons of Neptune. It orbits Neptune at a distance of 355,000 kilometers (213,000 miles) and it completes its orbit in six Earth days (one Tritonian month). It is approximately 2,705 kilometers (1,680 miles) in diameter, the tenth largest satellite in the Solar System, just slightly smaller than Earth's Moon. It was discovered in *CE 1846* by Lassell. Triton has an ice surface and where there is little evidence of cratering, there are numerous large cracks in Tritons ice surface, some as large as 200 kilometers (120 miles) across, evidence of vast geologic disturbances in the past, possibly a result of the Agent of Mass Destruction. Triton was named after the son of the god Poseidon in Greek Mythology. Triton revolves retrograde around Neptune. Most bodies in the Solar

System revolve in their orbits counter clockwise as seen from the north pole of the host, retrograde orbits revolve clockwise, in the opposite direction. There are very few retrograde orbits in the Solar System and this thesis proposes retrograde orbits are evidence of the Agent of Mass Destruction.

PROTEUS - Proteus is the second largest of Neptune's moons and it orbits Neptune at a distance of 118,000 kilometers (70,600 miles) and completes its orbit in just under twenty seven Earth hours (one Proteusian month). Proteus is irregular in shape and is just 436 kilometers (262 miles) in diameter at its widest point and 402 kilometers (241 miles) at its smallest point, about one ninth the diameter of Earth's Moon. It was found in *CE 1989* by Voyager 2. The irregular shape and size of Proteus, and its close relation to Saturn's moon Phoebe, along with the heavily cratered surface, indicate it may have started as an asteroid and undergone severe bombardment from some other celestial source during its history. Proteus was named after a god of the sea who could change his shape at will in Greek Mythology. Proteus is very similar to a moon of Saturn named Phoebe. They both have dark heavily cratered surfaces and although scientists propose both moons may have formed in the same region of space, because they are very similar, this thesis proposes they were once a single body broken in two, Phoebe being pulled inward by the Agent of Mass Destruction to be captured by Saturn and Proteus being left behind in the orbit of Neptune as it passed there.

NERIED - Neried is the third largest of Neptune's moons. It orbits Neptune at a distance of 5.51 million kilometers (3.33 million miles) completing an orbit in three hundred and sixty Earth days (one Neriedian month). It is 340 kilometers (204 miles) in diameter, about a tenth the size of

Earth's Moon. It was found in *CE 1949* by Gerard Kuiper. It has a highly cratered surface, evidence of numerous bombardments from space. The moon was named for a group of fifty sea nymphs who were daughters of the Greek god Nereus and his wife, Doris. Scientists propose Neried was once a comet from the Kuiper Belt, an area of origin for many of the comets of the Solar System, and in one of its passes toward the Sun, it was captured by Neptune. Whereas the icy nature of this small moon may indicate so, it may also be possible Neried was drawn to its present orbit by the Agent of Mass Destruction as it came marauding through the course of the planets.

LARISSA - Larissa is one of the small moons, the fifth closest of the inner moons of Neptune, egg shaped and measuring just 208 kilometers (129 miles) at its widest point and 178 kilometers (111 miles) at its narrowest. It could fit inside one of the larger craters on Earth's Moon. It resides in an orbit just 73,600 kilometers (45,700 miles) from Neptune and it completes its orbit in just thirteen Earth hours (one Larissan month). It was discovered in *CE 1981* by a team headed by the American astronomer Harold J. Reitsema born *CE 1948*. Its surface is black and heavily cratered showing evidence of past buffeting from space. Larissa may be a remnant of a past collision with the Agent of Mass Destruction, perhaps the same collision responsible for the rings and the expulsion of Pluto to the outer edges of the Solar System. Larissa was named for a character involved with Poseidon in Greek Mythology.

There are nine other moons of Neptune known at the time of this writing. There is little known about them, but they have been named. They are, in order of discovery, along with Proteus, Galatea, Despina, Naiad and Thalassa discovered in *CE 1989*, Halimede, Sao, Laomedeia and Neso discovered in *CE 2002*, Psamathe found in *CE 2003* and the last one unnamed merely designated as S/2004 N1 found in *CE 2013* by Mark R Showalter born *CE 1957*.

Pluto/Charon

Although scientists have recently demoted Pluto to a dwarf planet status, this thesis proposes including it among the planets because it has all the attributes of a legitimate planet. It is spheroid, has moons and as shown by New Horizon in *CE 2016*, it has an atmosphere, admittedly thin and wispy, but an atmosphere nonetheless. This thesis proposes when scientists revisit the question of Pluto, they will reinstate it to the family of planets.

Pluto is 2,320 kilometers (1,200 miles) in diameter, about two thirds the size of Earth's Moon. It takes slightly less than two hundred and forty eight Earth years for Pluto to complete one of its revolutions around the Sun (one Plutonian year). Pluto is 5.9 billion kilometers (3.67 billion miles) from the Sun.

Pluto was discovered in *CE 1930* by Clyde Tombaugh *CE 1906–1997* at the Lowell Observatory in Flagstaff, Arizona. Pluto was named for the god of the underworld in Roman Mythology, the equivalent to Hades in Greek Mythology.

Pluto, and its relatively large companion Charon, were once probably moons of Neptune, which became dislodged from its larger host by a close encounter with the Agent of Mass Destruction and were exiled to the outer reaches of the Solar System. It has a highly erratic orbit bringing it closer to the Sun than the orbit of Neptune at times, however there is practically no possibility Pluto and Neptune will ever collide since Pluto's orbit is inclined to the plane of the ecliptic by seventeen and two tenths degrees (17.2°). This means it orbits outside the plane in which all the other planets orbit. When it passes inside the orbit of Neptune, it only has a chance to encounter Neptune at two points, where their orbits intersect, and the chance of each of these planets being in the proper position for a collision at these times is extremely rare indeed, close to zero.

Pluto has five known moons, they are named Charon, Styx, Nix, Kerberos and Hydra, Charon being the largest. They were all named after mythological entities associated with Pluto, god of the underworld.

Charon is 1,270 kilometers (790 miles) in diameter, not quite half as large as Earth's Moon, making it the largest moon in the Solar System in relation to its host planet.

Charon and Pluto revolve around a common point always keeping the same faces to each other. They are 19,600 kilometers (12,200 miles) apart. They revolve around each other in six and a half Earth days (one Charonian month).

Charon was named after the boatman who ferries the dead across the River Styx into Hades in Greek Mythology. The name was chosen to compliment the name of the host planet, Pluto, god of the underworld.

The other moons have just recently been discovered by New Horizons and scientists are still gathering information about these far away moons as of this writing. I am positive, given the state of all the other bodies in the Solar System, planets, moons, rings, comets and asteroids, evidence of the Agent of Mass Destruction will also be found on the moons of Pluto.

Mythology

In Mythology, there is a brief description of what may have happened early in the history of the Solar System where Pluto was concerned. Pluto, in Mythology, assisted his brothers Neptune and Jupiter in the overthrow of Saturn, which left Jupiter as the head of the gods, Neptune as head of the sea, and banished Pluto to the underworld.

Scientists think Pluto was once a moon of Neptune, and if, in the history of the Solar System, Neptune and Jupiter came together in the vicinity of Saturn in such a way as to unseat Saturn from its place of preeminence in the Solar System, then the small moon circling Neptune at the time (Pluto), which assisted Jupiter and Neptune in the overthrow of Saturn, was flung out of its orbit by the encounter and was sent to the far reaches of the Solar System, the underworld.

Only an outside influence of gigantic magnitude could have precipitated such an encounter, an outside influence of great size, or one travelling at tremendous velocity, or both.

The fact there is this description of these celestial events handed down to us in the form of ancient tales and mythology is amazing, meaning the early inhabitants of planet Earth retained some knowledge of these events. This could only have come down as knowledge from God to the forefathers, or from the sons of God to early man, with whom they were intermingling, since they were in a primitive state themselves at these early ages in the Solar System, and certainly did not have the knowledge to devise the instruments necessary to directly observe these events as they occurred, because these events certainly occurred outside the range of direct vision of man.

Here is further evidence of the Agent of Mass Destruction, in the absence of any direct visual evidence a closer examination of Pluto and Charon is sure to reveal.

The Ages of the Earth

Prehistory

In the distant mists of prehistory, from the beginning to the destruction visited on the Earth by God at his sore displeasure at Satan's fall, even to after the Reconstruction but long before modern man came onto the scene, the Earth was inhabited by a vast assortment of plants and animals, most of which are now extinct. Everything from exotic carnivorous plants, to the wide array of dinosaurs roaming the Earth as masters of all they surveyed. The life span of these species was each considerably longer than the relatively short span of humankind upon this planet.

The dinosaurs walked the Earth for over three hundred million years. Man, by comparison, has existed here for about one million years, according to scientists.

The earliest skeleton, known as "Lucy", was the earliest ancestor of humankind thus far uncovered in the plains of Africa, being a mere one million years old, by the latest carbon dating method. This skeleton is of a creature more ape than man, upright in stature, and therefore leading quite inexorably, according to paleontological thought, to modern man.

Modern man has resided here for only fifty thousand years or so, and those of our ancestors with a marked intelligence less than fifteen millennium.

Eons, Eras and Periods

The Earth is roughly four and a half billion years old. For the first billion years, no life existed on the planet, it was forming the geology, cooling from a hot ball, as the scientists conjecture. Then around three and a half billion years ago, life began to appear, appearing and disappearing over the next three billion years in an endless opera of lifes tenuous hold on the surface of a hostile planet.

Many species came and went as the cycle of life tried desperately to take hold on Planet Earth. Then, about five hundred million years ago, life took hold and began to flourish on this planet, and displayed itself prominently in the drama which has unfolded, as scientists have surmised to this day based on the often ambiguous evidence..

These years are broken into Eons, Eras and Periods, as shown below.

mya = millions of years ago
Precambrian Age – (4,500 mya - 540 mya)
    Hadean Eon (4,500 – 3,800 mya)
    Archaean Eon (3,800 – 2,500 mya)
    Proterozoic Eon (2,500 - 540 mya)
        Paleoproterozoic Era (2,500 – 1,600 mya)
        Mesoproterozoic Era (1,600 - 900 mya)
        Neoproterozioc Period (900 - 600 mya)
        Vendian Period (600 - 540 mya)
    Phanerozoic Eon (540 mya - present)
        Paleozoic Era (540 - 245 mya)
            Cambrian Period (540 - 500 mya)
            Ordovician Period (500 - 445 mya)
            Silurian Period (445 - 408 mya)
            Devonian Period (408 - 360 mya)
            Carboniferous Period (360 - 280 mya)
            Permian Period (280 - 245 mya)
        Mesozoic Era (245 - 65 mya)
            Triassic Period (248 - 208 mya)
            Jurassic Period (208 - 146 mya)

Cretaceous Period (146 - 65 mya)
<u>Cenozoic Era</u> (65 mya - present)
Tertiary Period (65 - 1.8 mya)
Quaternary Period (1.8 mya - present)

Periods accompanying major mass extinction events are italicized for clarity.

Earth's Major Extinctions

Mass Extinctions

It is difficult to discern events in Earth's prehistory of a destructive nature, because the weathering factor on Earth, and the nature of erosion, wind, rain, volcanism and earthquake activity have all covered or erased many of the traces of these events. As one scours the Earth for clues, it is necessary to keep a discerning eye, and a keen intellect, to see the evidence. In the chapter on Earth geology, this study will attempt to show some of these events and the aftereffects of the events.

To see the periods of large destructive events on the Earth in a clearer light, it is necessary to look at the paleontological record, and notice events of mass extinction in Earth's prehistory.

There were many mass extinction events since the Earth has stabilized from the molten mass it was during its formation, since it has allowed life a foothold. During these mass extinction events major portions of the flora and fauna perished. There were also numerous events of extinction on a small scale, but for the purposes of this thesis, the concentration will be on those events with a high rate of decline to discern when this event may possibly have occurred.

There were probably many impact events in the Pre-Cambrian Time, before five hundred and forty million years ago, but life maintained such a tenuous hold and there is little or no evidence of any extinctions to bare it record, although there was some life, and there were undoubtedly extinctions, therefore it is not necessary to be concerned with the Pre-Cambrian Period in searching for this Agent of

Mass Destruction because the Agent caused the death of a planet and almost all the life on it.

Post Pre-Cambrian Extinction Events

Six of these events after the Pre-Cambrian Time are known as:

First in the Paleozoic Era, at the end of the Ordovician Period at the beginning of the Silurian Period (this study will call this the O-S boundary); approximately four hundred and forty five million years ago.

Second, in the Paleozoic Era, at the boundary of the Frasnian and Famennian Stages of the Late Devonian Period (the F-F boundary); approximately three hundred and fifty five million years ago.

Third, at the end of the Permian Period and the beginning of the Triassic Period, which also happened to be the end of the Paleozoic Era and the beginning of the Mesozoic Era (the P-T boundary); approximately two hundred and forty million years ago.

Fourth, in the Mesozoic Era, at the end of the Triassic Period and the beginning of the Jurassic Period (the T-J boundary); approximately two hundred and six million years ago.

Fifth, at the end of the Cretaceous Period and at the beginning of the Tertiary Period (the K-T boundary), this was also the end of the Mesozoic Era and the start of the Cenozoic Era; approximately sixty five million years ago.

And sixth, also in the Cenozoic Era, in the middle of the Tertiary Period, there was a rather large extinction at the end of the Eocene Stage and the beginning of the Oligocene Stage (the E-O boundary); approximately thirty three million years ago.

There are six major extinctions to choose from:

The O-S event.
The F-F event.
The P-T event.
The T-J event.
The K-T event.
The E-O event.

Scientists have established there are significant extinction events approximately every thirty million years, or so, and many species go missing during these events. The recovery periods in the intervals give rise to new species which in turn have their day at the extinction well during a later event. Man is merely the newest arrival on the scene in this ageless drama.

Major Mass Extinction Events

There were a number of extinction points in the vast time frame of Earth's prehistory before written history began to be recorded.

The Earth is about four and a half billion years old, considerably younger than the fifteen and a half billion years, or so, of the Universe. During this four and a half billion years, there has flourished a great diversity of life on this planet, and there is rich scientific evidence of this in the geologic and fossil record. There is also ample scientific evidence of specific extinction points in this immense period of time. There is evidence of extinction events roughly every twenty six million years or so, but there were six mass extinction events thus far uncovered in the impressive fossil record.

Ninety nine percent (99%) of all life which has flourished on this planet in the past, is now extinct, showing itself only in the fossil record.

Now for an examination of each of the major extinction events in turn from the earliest to the latest.

The O-S Event – 445 Million Years Ago

Approximately four hundred and forty five million years ago, there was a mass extinction on the Earth. The third greatest extinction event ever recorded.

The Ordovician period began about five hundred million years ago, at the end of the Cambrian, and ended around four hundred forty five million years ago, being followed by the Silurian. During this period, the northern half of the Earth was almost entirely ocean, and the land was concentrated in the southern part of the globe, making up the single landmass, the concentrated land mass spoken of in the Bible, **_Genesis 1:9;_** known by scientists as Gondwanaland. Throughout the Ordovician Period, Gondwanaland was shifting south, towards the South Pole, and much of it was covered with water.

During this period, life on Earth was defined by the invertebrates, the trilobites and brachiopods and by the early vertebrates, the conodonts. The ocean teemed with these invertebrates as well as corals and red and green algae, and some primitive fish, cephalopods, gastropods and crinoids. It is also possible land plants were flourishing at this time, possibly even beginning to invade the land.

Also during this period, the Earth's climate was milder, warmer with more moisture in the air. When Gondwanaland finally ended its migration to the South Pole, the cold settled in and glaciers formed, causing a drop in sea levels and the beginning of an ice age.

This mass extinction was not associated with an iridium spike, as are many of the others, and this one would appear to be caused by a major climatological change instead of an impact event. The first part of the extinction accompanied the beginning of the ice age, and the latter stage at its end. This event may end up being a relatively small one by comparison as more evidence is gathered. Scientists, at present, estimate approximately sixty percent (60%) of all marine life died during this extinction event.

Cambrian Extinction

The first of these mass extinction events occurred at the boundary of the Cambrian Period and the Ordovician Period, about five hundred million years ago. During the Cambrian, there was a great diversity of life which flourished, from the seas to the land. During this time the sea level was considerably higher, The land was marshy swampland, and the life which flourished on Earth at this time was mostly water borne, soft bodied, or soft-shelled at the least.

Because of the soft bodied constitution of these species, the fossil record is not very abundant with their residue, therefore making it harder for scientists to really understand the extent of the extinction.

When the extinction event occurred, nearly all these mollusks, echinoderms, brachiopods and trilobites went extinct. A few of the hardier species, those with a harder shell, survived and were able to diversify from there, because at this extinction event, the level of the sea was lowered allowing more land to appear, allowing these species to diversify toward a landward existence, becoming heartier still.

Since it is currently hard to tell the wide-ranging impact of this extinction event and the percentage of species involved, it will be mentioned but is not included among the six cited above.

Late Ordovician Extinction

After the Cambrian extinction, the remaining species began a worldwide diversification, growing and adapting to the sea and filling the Earth. During this period, the Ordovician, the Earth was extremely stable, allowing for a diversity of life to take root in the sea. At the end of this stable period about four hundred and forty five million years ago, at the boundary between the Ordovician Period and the Silurian Period, glaciation occurred on a grand scale, freezing much of the water into ice sheets, drastically lowering the sea level. Those species which heretofore were thriving in the sea, were unable to continue thriving because their living space, the sea, was severely decreased, and the salinization of the sea was greatly increased when the fresh water was captured into glaciers. A large number of these species came out of the water and adapted to life in the open air, on dry land. Then roughly one million years later, a warming period occurred, melting the glaciation, flooding the salinized water, (a salinization level to which most of the remaining sea bound species had previous to this adapted) with fresh water on a vast scale, thereby causing another period of extinction and a raising of the sea levels. The species which adapted to a life on the Earth, who stayed near the shores, which most earthbound life did, died out in the ensuing rise of ocean levels. Therefore about half of all species died.

These two events, relatively close together in geologic time, about a million years apart, are considered the second mass extinction at about four hundred and forty five million years ago.

The F-F Event – 355 Million Years Ago

In the Devonian Period, sea life was beginning to diversify toward the land in greater numbers. The fresh waters were beginning to teem with life and the amphibians and plants were flourishing on the land. At the barrier between the Middle Devonian and the Late Devonian, about three hundred and fifty five million years ago, another major extinction occurred, and there is much disagreement between scientists about the nature of this extinction, very little evidence existing. There is enough evidence to establish there was a major extinction of the sea life and it appears the fresh waters and the land creatures were not as heavily affected. About seventy percent (70%) of all sea life was wiped out during this extinction event.

Frasnian (Devonian) Extinctions

This event was heralded by an event of celestial importance because there was a rise in iridium in the layers associated with this F-F boundary. Also of note were shocked quartz and glass spherules found throughout the world, in China and in Europe during this time. This indicates an impact event, or perhaps a number of small impact events. A celestial Agent was involved, of this there can be no doubt.

However, there were also some meteorological changes occurring at this same time, which leads scientists to believe this mass extinction was in stages, some maybe caused by a celestial event and some by the resulting weather changes. It is believed Gondwanaland again drifted into a polar region and another ice age ensued.

It is also known the sea level rose and the chemistry of the ocean changed at the time of this extinction, thereby enforcing the theory of the end of an ice age. The rise in sea level, and change of the oceanic chemistry, caused the killing off of huge areas of coral reef on a worldwide basis. This in turn resulted in a depletion of the marine life dependent upon those reefs.

The fauna of this time was both marine and land dwelling and beginning to show more complexity than those is the Ordovician. During this event approximately forty five percent (45%) of the marine life went missing. This event was smaller in effect on the emerging land fauna.

The P-T Event – 240 Million Years Ago

The single largest extinction event in the long history of the Earth occurred at this time, 240 million years ago, at the boundary between the Permian Period and the Triassic Period.

During the Permian, and the period preceding it, the Carboniferous, life flourished on Earth. The plant life filled the land making it a tropical Eden, and animal life was diversifying at leaps and bounds, the amphibians and reptiles reaching out to the whole of the land. The seas were full, with a diversity of life more widespread than at any time in the past. The fish species were numerous as well as the simpler forms of shelled and soft-bodied mollusks. From three hundred and forty five million years ago to roughly two hundred and forty five million years ago, a period of one hundred million years, life was allowed to thrive without any major events to cause their demise, during a major period of climatic and geologic stability.

Then, for reasons not entirely agreed upon by experts, at the End of the Permian, there was a major extinction event, causing ninety nine percent (99%) of all life on the planet to perish.

This work ventures to express an opinion on the reason for this mass extinction, the Agent of Mass Destruction. It is this extinction event which this thesis proposes is the signpost in time to point to the fall of Lucifer, two hundred and forty million years ago, and the total destruction of the Earth at God's fierce anger.

Permian Extinction

According to Douglas Erwin, a noted American paleobiologist on faculty at the Santa Fe Institute and the Smithsonian Museum, studying mass extinctions, this one was the "Mother of all Mass Extinctions", the largest of all times.

John Phillips, *CE 1800-1874* a noted British geologist of his time, marked the end of the Paleozoic Era by this extinction event. Incidentally he also coined the term Mesozoic for the period following the Paleozoic.

After this event, life was completely different on Earth than ever before. Before this event, life mainly occupied the oceans and hugged the shores when it did venture onto the land. After this event, both areas flourished with life equally and the complexity of the land animals surged.

Approximately ninety nine percent (99%) of all species on Earth perished at this event, marine life as well as land animals.

Scientists are at a loss to explain what caused such a massive extinction at this time. There appears to be little evidence of an impact event, but many scientists think there was one. The geologic evidence of changes in the Earth at this time, though prolific, is not sufficient to explain an extinction of this magnitude. It may be possible there was a combination of events contributing to this extinction, or it may just be the Agent of Mass Destruction was the culprit.

I shall attempt to show the Agent was the catalyst of this event, and of all the geologic changes accompanying this event. This Agent will fit the bill as the offender.

The T-J Event – 206 Million Years Ago

At the end of the Triassic Period, about two hundred and ten million years ago, at the boundary between the Triassic Period and the Jurassic Period, there was another mass extinction, about twenty five percent (25%) of all species passing out of existence at this time. This made way for the rise of the dinosaurs during the Jurassic Period, plant life having flourished for millions of years, and animal life having diversified to the point of widespread occupation of the land space on Earth. The dinosaurs were set to move from a place of just another species vying for space on the Earth, to one taking charge for a hundred million years or so. Their turn would come.

Triassic Extinction

This was a small extinction event compared to those preceding it with only twenty five percent (25%) of the flora and fauna perishing. It is possible and noteworthy to explain it as another impact event. There are shocked quartz examples found in this layer in Italy and Canada and there are numerous impact craters throughout the world dating to this time. It appears as if a large body was on a path to encounter the Earth and fragmented into numerous smaller bolides as it entered the atmosphere, leaving much smaller craters as a result. The evidence is still being sifted and it is not certain this should be ranked as one of the major extinction events at all.

The K-T Event – 65 Million Years Ago

This is the most famous of all the extinction events, although certainly not the most devastating, like the distinction belonging to the P-T Event.

During this event, about sixty five million years ago, at the boundary between the Cretaceous Period and the Tertiary Period, nearly eighty five percent (85%) of all species died out, including the dinosaurs. This extinction made way for the ascendancy of the mammals, especially mankind, which for some reason, were hardly affected by the K-T event (K-T = Cretaceous/Tertiary).

At this time, something monumental destroyed the Earth, causing an end to the abundance of the preceding one hundred million years and nearly wiping out life on Earth entirely.

Scientists claim it was an impact by an asteroid in the Yucatan Peninsula of Mexico, and this fits well with the Agent of Mass Destruction.

Cretaceous Extinction

This is the one known for the end of the "Reign of the Dinosaurs". Although the dinosaurs, alive at the time of this impact, perished, it was, in fact, not the end of the age of the dinosaurs, as many other species of what would be termed dinosaurs came to the fore afterward.

Scientists now know it was caused by an impact event at the tip of the Yucatan Peninsula in Mexico, at Chicxulub. They estimate the object impacting there was approximately 10 kilometers (6 miles) in diameter and left a crater over 160 kilometers (100 miles) in diameter.

It is estimated eighty five percent (85%) of all species perished during this event, making it the second largest of the extinction events, only surpassed by the P-T boundary event.

The E-O Event – 33 Million Years Ago

This event was also relatively minor by comparison. This event affected marine life more than land animals and it was not caused by any impact event yet detected. There are not the usual rises in iridium levels and the shocked quartz evidence usually accompanying an impact event. There also appears to be no major volcanic eruptions at this time, and this also can be ruled out as a cause of this event.

Approximately twenty percent (20%) of the marine life died as a result of this event.

Summary

There actually seems to be a pattern of extinction events occurring approximately every thirty to sixty million years. The last event was at thirty three million years ago, preceded by the big one at sixty five million years ago, when the mighty dinosaurs, who ruled the Earth for over one hundred million years, became extinct.

Prior to this there were minor events at 130 million years and at two hundred and six million years. Then our P-T event at two hundred and forty million years, preceded by a smaller event at three hundred and ten million years.

Then came the last two major events at three hundred and fifty five million years, and four hundred and forty five million years. Before this, the Earth was still forming and molten in most areas, and extinctions are hard to find.

China's Carbon Isotope Spikes

In *CE 1998* scientists reported new data from China. They found the carbon isotope change at the E-O boundary of thirty three million years ago was probably very short-lived: only a "spike", perhaps one hundred and sixty five thousand years long at the most. This suggests a major (catastrophic?) addition of non-organic carbon to the ocean, rather than just a failure in the supply of organic carbon.

They suggested three possible scenarios. Two of them are variants of the Siberian Traps scenario, except in addition, the climatic changes could have set off an overturn of Panthalassa and a carbon dioxide crisis. Their third suggestion is an asteroid impact, but scientists are still looking for evidence of this impact.

Smaller Extinction Events

In addition to the very largest mass extinctions, American paleontologists David M. Raup *CE 1933–2015* and J. John Sepkoski Jr. *CE 1948-1999* identified smaller ones. They noticed mass extinctions seemed to have occurred periodically, every twenty six million and two hundred thousand years since, at least, the end of the Permian. The suggestion was followed immediately by a flurry of claims and counterclaims about cycles in the geological record. Others suggested the extinction events occur in a thirty million year periodicity, or the extinctions matched a periodicity of twenty eight million and four hundred thousand years for impact craters on Earth's surface, or there were periodicities between thirty and thirty four million years for crater ages, magnetic field reversals, plume eruptions, pulses of mountain-building, and other events.

Catastrophism

In Earth's past, and throughout the world are numerous and stark evidence of catastrophic events which helped mold the planet. Life, in all forms, is sometimes the direct result of this catastrophism, both the birth of the life and the death of same, that is why it was necessary to look at these events in more depth.

Earth Changes

In the record of the Earth, there is much evidence for catastrophe, even though scientists go out of their way to ignore this evidence. Scientists are vehemently set against a catastrophic scenario to the extent they have developed elaborate and complicated theories about the origin and evolution of an ecosystem, theories in one science (Biology) which violates the rules of another discipline (Physics).

The Theory of Evolution

In Biology, the Theory of Evolution, *CE 1859* was popularized by the release of "The Origin of the Species" by renowned author and controversial English naturalist

Charles Darwin *CE 1809–1882*. In the simplest and most debatable terms, it controversially declares man evolved from one less developed creature to a more advanced form, on and on up, all the way from amoeba to man. This was all done by natural selection through the process of survival of the fittest, nature preferring one form over another, eventually leading to man. All this was done in a slow and steady process, without any outside influences, over millions of years, slowly, ever so slowly. They refer to it as geologic time.

The Second Law of Thermodynamics

In Physics, the Second Law of Thermodynamics, discovered in *CE 1855* by Rudolf Clausius *CE 1822–1888*, establishes the total entropy of an isolated system always increases over time, or remains constant in ideal cases where the system is in a steady state or undergoing a reversible process. The increase in entropy accounts for the irreversibility of natural processes, and the asymmetry between future and past.

To experience entropy means to deteriorate.

Just as scientists in Biology state their laws of evolution are immutable, Their <u>Theory</u> of Evolution appeared four years after the Second Law of Thermodynamics were established. The scientists in Physics state their laws are immutable, also. It isn't a matter of who was first, it is a matter of who is right. The Laws of Physics are immutable. The Theory of Evolution is not a scientific law. Physics rules. This is not to say the Theory of Evolution is incorrect, it merely points out it is a theory and if this theory violates a law, it comes into question.

Catastrophism bridges the gap, being an external agent which could affect a system causing an organism to move up the evolutionary chain, if it added the right ingredients to the system.

Scientists are like an ostrich with his head in the sand, ignoring the evidence rather than admit the possibility all the questions their theories present are explained in a catastrophic scenario. Why some intransigent scientists are against catastrophe as a means to accomplish our history is beyond reason, but they do, nonetheless, *"kick against the pricks"* where widespread catastrophism is concerned. A catastrophic scenario in Earth's prehistory answers so many questions currently being raised by the slow and steady geologic process scientists currently favor. The recent assent of some scientists to the asteroid impact theory at Chicxulub, Mexico as a means of the extinction of the dinosaurs and the rise of the mammals is a major step in the right direction toward the acceptance of catastrophism as a cause of the other major extinction events.

Now to look at some of the evidence on the Earth speaking to this issue. There is both geologic evidence and biologic evidence (paleontological evidence), and a look at both will be revealing, beginning with the geologic record.

Geologic Evidence on the Continents

First, it is necessary to examine the geologic evidence, which is plentiful throughout the Earth.

There are many places on the Earth where lava has poured out onto the continents, without the agent of a volcano, this is known as continental rifting and outflow. It is basaltic rock melted to a molten mass and in mile wide fissures has seeped across millions of square miles in many places. It is necessary to investigate a few of these areas by continent.

Asia

Our largest continent holds many anomalous geologic areas that demands attention.

The Siberian Traps

The Siberian Traps were formed by outflow of molten material onto an area covering over 4 million square kilometers (2.5 million square miles) in central Russia, in Siberia, sometime near the end of the Permian period. The reason for its existence is in debate, some scientists saying there is not enough iridium evidence for an impact scenario at this time.

This thesis shows why there was an impact, an impact larger than any before or since, at this time, and why this impact left little or no iridium evidence. The evidence is the large scraping of soil and rock down the whole length of the North American continent.

Realize, scientists also said the same thing for many years about the K-T event (Cretaceous/Tertiary), before they found an impact crater in the Yucatan and a 7.5

centimeter (3 inch) thick layer of iridium in all soil strata layers at this precise moment in time. The very fact an impact crater was found in the Yucatan to explain the mass extinction at the K-T boundary, now leads scientists to begin looking harder for impact craters to accompany all mass extinction events, and they think they now have found one for the P-T boundary (Permian/Triassic).

A team of scientists led by American geologist Dr. Luann Becker of the University of California, Santa Barbara, doing research may have found definite evidence of an impact event at the P-T boundary, some two hundred forty million years ago. The team found some complex carbon molecules called fullerites which show a high concentration at the time of the Permian mass extinction, which indicates an impact.

Also, a team of other researchers in *CE 1996* found evidence of impact at the P-T boundary in fungal cells found in the sediments indicating an increased amount of fungi breaking down massive amounts of vegetation at this time. This would indicate a massive annihilation of plant life at this time which also indicates an impact.

The Deccan Highlands

The Deccan Highlands of India covers almost the whole of the interior of the sub-continent constituting the whole of the south Indian tableland, the elevated region lying east of the Western Ghats. This outflow area is antipodal to the impact site at Chicxulub, Mexico. It rises to over 1,000 meters (3,280 feet) in places and is eroded by river valleys throughout.

The causes of the Deccan Highlands are in debate, as are the Siberian Traps and the Guyana Massif, all similar in nature.

The Deccan Highlands is an area of outflow of molten rock at some point in prehistory, (probably at the same time as the K-T boundary impact event at the Yucatan Peninsula in Mexico), which flowed over the area creating the Highlands. As previously shown with the Siberian Traps, this was the result of an impact at the other end of the Earth. At the time of the K-T event, India was still moving into its present position after the breakup of Pangaea, and occupied the exact opposite side of the Earth from the Yucatan, thereby solving this dilemma.

Impact Craters

There are twenty nine known impact craters on the Asian continent. Considering the vast extent of the Asian continent, this is a small number of craters, indeed. With all the sprawling kilometers exposed to the ravages of space, this continent should have many more, just by the law of averages.

The oldest known crater in Asia is called the Suavjarvi Crater in Northwestern Russia. It measures 16 kilometers (10 miles) in diameter and was formed in the Mesozoic Era in the Triassic Period around two hundred and four billion years ago.

The largest known crater in Asia is called the Puchezh-Katunki Crater located in Russia in the area of the Volga River. It measures 80 kilometers (50 miles) in diameter and was formed approximately one hundred and sixty five million years ago, also in the Mesozoic Era, except this one was in the Jurassic Period, when dinosaurs roamed the Earth.

The youngest and smallest is known as the Sikhote-Alin Crater in Southeastern Siberian Russia. It measures only 26 meters (85 feet) in diameter and was formed in *CE 1947*.

Europe

The Flattened Mountain in Austria

In Austria, there is a mountain in the Southern Alps, near Köfels, which, at some time in the past, near *3100 BCE*, appears to have collided with an asteroid, supposedly an Aten type asteroid which struck the Earth around the same time, and the mountain lost its top and much of its side. The top of this mountain is craggy and much shorter than the accompanying mountains. There is a mountain tarn at the level part near the top where a large area was carved out and filled with water.

Some scientists are hypothesizing this event occurred based on a Mesopotamian circular stone tablet depicting the passage of a meteor across the sky and they are saying the meteor caused this event, called the Köfels Impact Event. The ancient artifact dates to about *3100 BCE* and was partially destroyed during the sack of Nineveh.

Others say this comet hit this mountain and the blowback from the impact threw material high into the sky, only to rain back down as fiery bolides on the Middle East, particularly Sodom and Gomorrah.

In my studies, Sodom and Gomorrah happened thousands of years after this. This thesis proposes this Sodom and Gomorrah interpretation is in error, however, this very well could be more evidence of the Agent of Mass Destruction.

Impact Craters

In Europe there are forty one known impact craters throughout the continent, having a wide variety of diameters and man has, in many cases, inhabited the centers of many of these craters in his need for space. Europe has the distinction of being in second place among the continents in number of impact craters found. This is probably a result of how long it has been populated by inquisitive minds, after all, almost all scientific advancement in the *CE Nineteenth and Twentieth Centuries*, the centuries witnessing the greatest expansion of exploration of a scientific nature in history, originated in Europe.

One of the most prominent of these is the Nordlinger-Ries in Bavaria in Germany measuring over 24 kilometers (15 miles) in diameter and has a few towns inside the crater, Nordlinger being the largest with Monchsdegginen being also worthy of mention. Approximately 42 kilometers (26 miles) to the west southwest is another crater, the Steinheim crater, measuring 3.8 kilometers (2.36 miles) in diameter which is believed to have formed at the same time as the Nordlinger-Ries. They surmise an asteroid broke into two pieces and landed separately on the area about fourteen million years ago.

The most recent and the smallest is the Kaali Crater in Estonia which formed between *3000 to 1000 BCE* and is approximately 110 meters (360 feet) in diameter. There are nine craters in the area of which Kaali is the largest and they were presumably formed by a disintegrating bolide entering the Earth's atmosphere.

The largest is the Siljan Ring an impact crater in Dalarma, Sweden. It measures 52 kilometers (32 miles) across and was formed approximately three hundred and seventy five million years ago.

The oldest impact zone currently measured is the Paasselka Crater in Finland and the Keurusselka Crater, each with an age of approximately one billion, eight hundred nillion years old. The Paasselka crater measures 10 kilometers (6 miles) in diameter. The Keurusselka Crater is 30 kilometers (19 miles) across.

There are two vying for the distinction of being the next oldest, both nearly one billion years old. They are the Lumparn Crater in Finland and the Suvasvesi Crater, also in Finland. The Lumparn Crater measures 9 kilometers (5.5 miles) across and the Suvasvesi Crater is 4 kilometers (2.5 miles) across. As the continent of Antarctica gives up more of its secrets, there may prove to be some there older than all three of these Finnish craters.

These could very well be pieces from the Agent of Mass Destruction, as could all the other many craters throughout Europe.

South America

South America is replete with many geologic features which bespeak catastrophism.

The Falkland Plateau

Antipodal to the Siberian Traps is the Falkland Plateau. Scientists have now found evidence of the actual impact site for this event at the Falkland Plateau off Tierra del Fuego, South America, which just happens to be the other side of the Earth going straight through from the Siberian Traps.

On this submerged plateau, Michael Rampino of the Goddard Institute of Space Studies and New York University found two circular basins of larger size than at

Chicxulub, Mexico, indicating a larger impact there than at the K-T boundary. Rampino dated the rocks there to the time of the late Permian, our P-T event. The Siberian Traps also date to this same time.

This is also very similar to another impact, the K-T event, which caused a similar outflowing scenario at the Deccan Highlands in India at a later time, at the end of the Cretaceous, the K-T boundary.

The Venezuelan Tablelands (The Guyana Massif)

Another area of vast molten outflow, at a time even more distant than the time of the P-T event are the Tablelands of Venezuela, called "Tepuis". This is a vast area of uplifted igneous rock, forming a plateau at high altitude, which over the eons was heavily scored by erosion. It is tremendously older than the Siberian Traps or the Deccan Highlands and therefore date to an earlier catastrophe, perhaps the one from the Devonian Period, or even the Ordovician Period. The outflow of molten material was as large as the Siberian Traps, and after the outflow, there was an uplifting of the whole area, which would indicate a massive collision on the other side of the globe, pushing the Tablelands upward to the height of a few thousand feet. The uplifting event created an even greater impact on the Andes Mountains, uplifting them to their present heights and with their sharply inclined surfaces. The Tableland uplift was more uniform, and therefore the top of the Tablelands are flat and level.

An event of this magnitude would necessarily be accompanied by another large extinction, which leads me to believe it was the one of the Ordovician Period.

In Venezuela, the highest waterfall in the world, Angel Falls, cascades over 979 meters (3,212 feet) from the top of the uplifted Tablelands to the jungle floor below.

The largest lake in South America, Lake Maracaibo, over 13,210 square kilometers (5,100 square miles) in area, was formed in the distant past when a massive draining of water off the land collected in this basin, whose origin is in question. Lake Maracaibo is in the form of a teardrop, and at its bottom are huge reserves of petroleum, which indicate an extraterrestrial source as discussed in the section on Hydrocarbons. The Lake Maracaibo basinm according to some scientists, was formed by an impact.

Also, the third longest river in South America, The Orinoco, which originates at over 305 meters (1,000 feet) above sea level in the Guyana Highlands (The Tablelands) flows north and east to the Caribbean Sea for approximately 2,560 kilometers (1,590 miles).

Also in Venezuela are some of the most ancient rocks on Earth, The Guyana Massif, an uplifted molten flow of millions of square kilometers in area, part of a Pre-Cambrian outpouring believed to be about two billion years old. This would place its formation in the Paleoproterozoic Era, in the Proterozoic Eon, when the Earth was just about finished cooling from its turbulent youth. It covers most of northern South America from Colombia, across southern Venezuela, into Guyana, French Guiana and Suriname and into Brazil. In Venezuela, the Tablelands are a part of this Massif.

A Sheared Mountaintop in Nazca

In the area of Peru known as the Nazca Plateau, there is a mountain appearing to have the top of the mountain flattened into what appears to some to be a runway, like at an airport. Since the whole plain of this area is replete with petroglyphs and arrow straight lines etched into the rocky flats by moving the dark rocks on top of the ground exposing the lighter underlayment. The detritus of the lines and the figures carved into the plain is there beside each excavation, but the detritus of the sheared mountaintop is nowhere to be seen. If man removed the mountaintop, the remains of his labor would be filling the valleys below the mountain. There is no evidence of man's activity in this phenomena. It is as if a big knife came slicing through there and cut the top off the mountain.

Perhaps, like the Ungava Peninsula in North America, or the mountain in Austria, this may be evidence of the Agent of Mass Destruction, scraping close to the Earth at one of its numerous passes in the past.

Impact Craters

There are fifteen known impact craters on the continent of South America, the least of any continent except Antarctica, which has gone largely unexplored because of the icecap and the harsh weather conditions there.

South America is relatively sparsely inhabited except for coastal areas. This may be why relatively few craters have been discovered on this continent. Also, where craters may have formed in the past, in the vast interior, a rain forest, the largest on the planet, would tend to erase evidence of impacts and hinder man's attempts for exploration to these areas.

The oldest crater in South America is the Serra da Calgahla Crater in Brazil. It measures 12 kilometers (7.4 miles) in diameter and was formed in the Paleozoic Era approximately three hundred million years ago, during the Carboniferous Period.

The largest crater is known as the Vichada Structure in Columbia, measuring 50 kilometers (31 miles) in diameter and having been formed about thirty million years ago in the Cenozoic Era.

As is very often the case with impact craters, the youngest is also the smallest. In the case of South America, this distinction goes to the Carancas Crater in Peru. It is a mere 13.5 meters (44 feet) in diameter and was formed very recent indeed, in the year *CE 2007*.

As seen on almost every continent, the older craters are in many cases the largest also, and the younger ones are also the smallest. This confirms the premise the earlier encounters with the Agent of Mass Destruction were larger chunks of rock than later ones have been, until these days small pieces are all which remains of the interloper.

North America

North America, like its southern neighbor suffered much catastrophism throughout its formation.

The Columbia Plateau of the Pacific Northwest

In the Pacific Northwest of the United States of America, around the Columbia River Basin, are basaltic lava flows forming the Columbia Plateau. These lava outflows, or basaltic floods, occurred between seventeen and twelve million years ago from the late Miocene Era to the early Pliocene Era.

These outflows erupted from cracks in the Earth several kilometers long, which covered an area over 170,000 square kilometers (41,000 square miles) wide. Of the two hundred and seventy lava flows spreading over this area, twenty one poured through the Columbia River Gorge, building the Plateau to a height of 1,800 meters (6,000 feet). The Gorge cuts through this strata to a depth of over 600 meters (2,000 feet). As the lava continued to erupt into the Basin, it flowed toward the sea in the same course now used by the Columbia River. These incessant lava outpourings filled many stream valleys and created dams leading to lakes in their wake. Many fossilized plants and animals are now found in the lake deposits. When the outpouring of lava stopped, the weight of the plateau began to warp the crust beneath it causing an up-thrust of the northern mountains. This caused the plateau to tilt slightly toward the south. This tilting played an important role in the development of the Grand Coulee.

This area in the Pacific Northwest is antipodal to Western Australia where numerous impact sites of this age are present.

The Yellowstone Plume Event

Plume events are rare: there were only eight enormous plume eruptions in the last two hundred and fifty million years. The most recent is the Yellowstone plume.

About seventeen million years ago it burned through the crust to form enormous lava fields known as the Columbia Plateau basalts of Oregon and Washington, best seen in the Columbia River gorge. North America drifted westward over this "hot spot," which continued to erupt to form the volcanic rocks of the Snake River plain in Idaho (Valley of the Moon, etc.), and it now sits under Yellowstone National Park. The hot spot is in a quiet phase now, with geyser activity rather than active eruptions, however, it produced enormous volcanic explosions about six hundred forty thousand years ago and blasted ash over most of the mountain states and into Canada. There is a periodicity to these eruptions of about six hundred thousand years and the last one was six hundred and forty thousand years ago.

North America is overdue.

Stone Mountain, Georgia

Although a much smaller outflow, and one pretty much covered over by soil deposits over the millennia, Stone Mountain, Georgia may be another area of outflow caused by an extraterrestrial Agent. At Stone Mountain, there is a bubble of granite extruding from the ground and rising into the sky. For hundreds of miles around the mountain, just under the ground surface, is a massive granite substrata. Anyone who has ever dug very far into the soil of Northern Georgia has encountered this outflow area. Although this is a much smaller outflow area than the millions of square miles of the Deccan Highlands, the Siberian Traps, the Venezuelan Tablelands or the Columbian Plateau, it is nonetheless proof of a smaller collision event sometime after the Tertiary Period.

The North American Continent and Some Anomalous Landscapes

This study has examined some of the evidence of the outflows, the massive basaltic flows having occurred on the Earth in the past. There is also more evidence of cataclysmic encounters between the Earth and other celestial agents. It is now time to look at some of this evidence.

The Hudson Bay

Hudson Bay, founded *CE 1607*, was named after its European discoverer the English explorer Sir Henry Hudson *CE ca. 1565-1611* when he was commissioned by the Muscovy Company to find a Northwest Passage. It is also historically named Hudson's Bay although the native population, the First Nation inhabitants of the area, the Eastern Cree, call it Winipakw, from which the city of Winnipeg, Manitoba derives its name.

Hudson Bay covers an area 1.2 million square kilometers (470,000 square miles) in extent. It drains most of Canada and even some of the United States east of the Rocky Mountains. It is fairly shallow for its size and position, only 100 meters (330 feet) deep and this is unusual for the area.

This thesis proposes it was formed by the same Agent which later scraped across the North American continent. This was the initial contact point for an event which will shortly be described as this study follows the track of the intruder.

The Ungava Peninsula

Whereas the Ungava Peninsula appears, at first glance to be another basaltic lava flow of the Siberian Traps and Deccan Highland type, it is however not of this type. Instead, it is an exposure of the bedrock of the area, bereft of any topsoil.

This is evidence for an encounter of Earth with an extremely large celestial body, which scraped the Earth at the Ungava Peninsula in northern Quebec, and left a cratered and fractured landscape all across the middle and western United States at some time later than the P-T event. The body encountering the Earth at this time scraped the ground of the Ungava Peninsula down to the bedrock, clearing away the topsoil completely, and to this day, there is little to no topsoil left in this vast expanse. All across this exposed bedrock are striations miles long bespeaking a scraping action by some large object.

Scientists have surmised the ice sheet, which covered the Northern Hemisphere around fifteen thousand years ago created this exposed rock with all its striations as it retreated. If the ice sheet caused this area to be exposed down to the bedrock, why wasn't the whole area, formerly covered by the massive ice sheet, as far south as Ohio, uncovered down to the bedrock when it retreated. There was some other Agent responsible for the exposing of the bedrock in the Ungava Peninsula and for the striations in the rock once exposed. This was the result of an encounter between the Earth and the Agent of Mass Destruction.

The Badlands of the South Dakota

Then, after this Agent of Mass Destruction scraped the ground at the Ungava Peninsula, the large celestial body proceeded on its trajectory and scraped the Earth again, though not as deeply this time, leaving in its wake the jagged Badlands of South Dakota and Wyoming.

The Canyonlands of Utah

Then it proceeded further along its path forming and shaping the Canyonlands of Utah, skipping slightly away from the surface after it formed the canyon lands, like a stone on the water, hitting the Earth again in Arizona where it cleft the Earth leaving the Grand Canyon in its wake.

The Grand Canyon

The fact scientists think the Grand Canyon was formed by erosion from the Colorado River over millions of years is understandable in light of the fact the Colorado River flows at the bottom of the canyon. This would be true no matter how the canyon formed. At the rim of the Grand Canyon, the effect is one of the Earth actually splitting apart at this point, a stress fracture, rather than a slow erosional process, and this is exactly what appears at the Grand Canyon, a stress fracture caused by the Agent of Mass Destruction.

The Baja Peninsula and the Gulf of California

The Agent of Mass Destruction then proceeded to scrape the Baja in Western Mexico and gouged out the area now serving as the Gulf of California. This planet sized body left much damage in its wake, leaving a landscape which looked ragged and mysterious, from the striations in the rock at the Ungava Peninsula to the craggy Badlands, to the obstruct laden Canyonlands, to the jagged rift of the Grand Canyon to the flat and inhospitable Baja.

Then the marauder left the Earth and headed back out to space, lingering in the vicinity of Earth to do more damage in the near future. And when it finally departed the inner Solar System, if it did depart, it left many large and small remnants to wreak havoc on Earth for millennia.

Impact Craters

There are sixty known impact craters in North America. Though North America is the youngest of the continents, it seems to have the most craters. This is probably due to the scientific interest in geology, geography, archaeology and anthropology encouraged in the scientific pursuits from the American universities and by the United States government. Since the early *CE Twentieth Century*, the United States has led the world in scientific endeavors, just as Europe did in the two centuries previous. It is not uncommon the youngest continent should exhibit the most impact craters. It is not because North America was hit more often than any other continent, it is because it has been more widely explored than any other, and in a much shorter period of time.

The oldest crater and also the largest is the Sudbury Crater in Ontario, Canada. It measures 250 kilometers (155 miles) in diameter and was formed approximately one billion, eight hundred million years ago in the Neoproterozioc Period.

The youngest and smallest of the North American Craters is the Whitecourt Crater located in Alberta, Canada. It measures 40 meters (131 feet) in diameter and was formed approximately eleven thousand years ago when civilized humans were just beginning to collect into settlements in North America.

Africa

Africa also shows much evidence of catastrophism during its long history.

The Karroo Basalts of South Africa

In South Africa, there are areas of basalt known as the Karroo System. It stretches from the north where the flows are thinner than in the south, only 1 kilometer (3,280 feet) in thickness and sometimes thin to the point of non-existence. In the south, near the Cape, the flows are 7.5 kilometers (4.7 miles) thick. These flows proliferated from the arid Triassic Period to the early Jurassic Period (two hundred million to one hundred ninety million years ago).

This area is antipodal to the Ungava Peninsula of Northern Canada where it was shown a major contact event occurred at about this same time.

The Great Rift Valley

The East African Rift System is part of a 5,000 kilometer (3,100 mile) fracture zone extending from the Limpopo valley in Botswana at the southern extreme to the Jordan valley in Israel/Syria at the northern extreme.

It came into existence in its southern part in the late Mesozoic Era, about one hundred million years ago. It was associated with voluminous igneous activity. Further extensive igneous activity throughout the Tertiary Period to include recent volcanos in the northern part of the rift. Fault movement forming the rift valley took place mainly in Miocene and Pleistocene times, and the area is still marked by seismic activity, volcanism, and high heat flow through the crust. Again this corresponds to the Ungava Peninsula event.

The Red Sea between the Saudi Peninsula and Africa is part of the Great Rift System. During Oligocene and Miocene times a new marine trough was developed on the site of the present Red Sea. Clays, marls, evaporites, and limestones accumulated in Miocene and Pliocene times, and then the two sides began to split apart as new sea floor was generated at the floor of the trough. The Red Sea therefore seems to be a new ocean basin in the making.

The feature runs north to south for over 5,000 kilometers (3,100 miles), from northern Syria to Mozambique and Botswana. The valley varies in width from 30-100 kilometers (19-62 miles) and in depth from 200 to 2000 meters (650 to 6,560 feet).

The northernmost part of the Rift forms the valley of the Jordan River, which flows southward from the Sea of Galilee to the Dead Sea. From the Dead Sea southwards, the Rift is occupied by the Wadi Arabah and then the Gulf of Aqaba and the Red Sea.

In eastern Africa the valley splits into two, the Eastern Rift and the Western Rift.

The Western Rift contains some of the deepest lakes in the world, up to 1,470 meters (4,823 feet) deep at Lake Tanganyika. Lake Victoria, the second largest lake in the world, is considered part of the Rift Valley system although it actually lies between the two branches. It is possible Lake Victoria may have been formed by an impact, if so, it would be the largest impact ever recorded, beside the conjectured Hudson Bay impact.

In Kenya the valley is deepest to the north of Nairobi. As the valley has no outlet to the sea, its lakes tend to be shallow and have a high mineral content as the evaporation of water leaves the salts behind. For example,

Lake Magadi is almost solid soda (sodium carbonate), and Lake Elmenteita, Lake Baringo, Lake Bogoria, and Lake Nakuru are all strongly alkaline, while Lake Naivasha needs to be supplied by freshwater springs to support its biological variety.

The formation of the Rift Valley is currently ongoing. In a few million years, eastern Africa will probably split off to form a new landmass, as did Madagascar in long ago ages. The original activity causing the Rift weakened the Earth's crust. The area is therefore volcanically and seismically active, producing the volcanoes Mount Kilimanjaro, Mount Kenya, Mount Meru, and Mount Elgon, and the Crater Highlands in Tanzania.

The Rift Valley is a rich source of anthropological discovery, especially in Olduvai Gorge in Tanzania. The bones of several hominid ancestors of modern humans were found there, including those of "Lucy", a nearly complete australopithecine skeleton, which was discovered by the American paleoanthropologist Donald Johanson born *CE 1943*. Also, Kenyan paleoanthropologist Richard Leakey born *CE 1944* and his wife American paleoanthropologist Maeve Leakey born *CE 1942* have also done significant work in the Lake Takana region.

The Rift Valley was most probably formed, at its beginning, by an encounter with some large celestial object, a planet sized body striking the Earth and splitting it open. Since the Great Rift Valley is antipodal to parts of North America, it could easily be the result of the encounter described above, scraping the Ungava Peninsula and creating the Badlands. This collision of the Earth with the Agent of Mass Destruction could easily have caused the fracture on the other side of the planet today recognized as the Great Rift Valley.

The Dykwa Glaciation Sequence

Towards the end of the Upper Palaeozoic new basins of deposition were initiated within the continent of Africa and on its margins. This phase began with a widespread glaciation in Carboniferous times. Huge thicknesses of glacial deposits called tillites, which are varved clays, and sandstones, were laid down; the Dykwa glacial sequence of the Karroo basin is 800-900 meters (2,600-2,950 feet) thick, suggesting a lengthy glaciation.

Sedimentation in the continental Karroo basin continued without interruption through Permian, Triassic, and Jurassic times. The Permian glacial deposits were closely followed by the deposition of coal measures (of economic importance in the Wankie basin of Zimbabwe), followed in turn by shales, sandstones, and finally the Karroo basalts. Large volumes of plateau basalts, up to 1000 meters (3,280 feet) thick, poured out in late Triassic and Jurassic times, and marked the beginning of the splitting up of the continent of Gondwanaland.

Impact Craters

There are nineteen known impact craters on the continent of Africa. For being the oldest continent on Earth, it seems to have very few craters, the least being South America, aside from Antarctica, which, as previously mentioned, has gone largely unexplored and there are probably many more craters hiding there under the ice.

Africa also may produce many more craters as this pursuit gains adherents the world over. Since crater hunting has become popular, since Gene Shoemaker *CE 1928 – 1997* popularized it in geological circles, Africa has gone largely unexplored because of all the political unrest throughout the continent.

The oldest crater on the African continent is known as Vredefort, formed in South Africa approximately two billion years ago in the Pre-Cambrian Period just after Earth's molten state cooled enough to leave evidence on the surface of the impact. It measures 300 kilometers (186 miles) in diameter making it claim the distinction of being the largest in Africa as well.

The youngest crater in Africa, and also the smallest is known as Kamil and is located in Egypt. It measures a mere 45 meters (148 feet) in diameter and was formed only about five thousand years ago, about the time of the Early Dynastic time in Egypt when civilization was beginning to flourish in the Nile Delta, before any of the pyramids were built.

Africa's craters again exhibit the pattern of the oldest being the largest and the youngest being the smallest.

Australia

Impact Craters

There are twenty seven known impact craters on the continent of Australia.

Australia, like Europe, is literally pock-marked with impact craters from throughout geologic history. There are only a few craters yet discovered because, like South America, Australia is a continent whose population inhabits the coastal areas and the inner reaches (the Outback) have gone largely unexplored. As more of the interior is explored with a specific interest in impact zones, it will probably produce many more craters to add to the list.

The largest confirmed crater is the Acraman Crater in Southern Australia which was formed approximately five hundred and ninety million years ago. It measures 90 kilometers (56 miles) in diameter.

The Woodleigh Crater, in Western Australia, was formed about one hundred and sixty million years ago. The crater is not exposed at the surface and therefore its size is uncertain. The original discovery team believe it may be up to 120 kilometers (75 miles) in diameter, but others argue it may be much smaller, with one study suggesting a diameter closer to 60 kilometers (37 miles). The larger estimate of 120 kilometers (75 miles), if correct, would make this crater tied for the fourth largest confirmed impact structure in the world, and imply a bolide (asteroid or comet) about 5–6 kilometers (3.1–3.7 miles) in diameter. A more recent study suggests the crater could be between 60 and 160 kilometers (37 and 100 miles) or more, and was produced by a comet or asteroid 6 to 12 kilometers (3.5 to 7 miles) wide.

The most recent and the smallest crater was the Dalgaranga Crater which was formed in about *1000 BCE* in Western Australia, approximately 120 kilometers (75 miles) west of Mount Magnet. It measures only 20 meters (65 feet) across which is quite small by astronomical standards.

When comparing the Dalgaranga Crater with the Kaali Crater in Estonia, they appear to be similar in size, nature, frequency and age.

If one studies impact Craters they can easily see earlier impacts happened alone and by large bolides. Then as time progressed, they became smaller and began breaking up as they fell. As one gets nearer and nearer to the present, these craters become fewer and smaller than any forming before.

This indicates a very large body started interacting with the Earth early on and left its trail across the surface of our planet, gradually disappearing from our records. However, the search for NEOs (Near Earth Objects) estimates there are over ten thousand NEOs currently being observed.

It seems the Agent of Mass Destruction has diminished in size and potency as man's history has unfolded.

Antarctica

Impact Craters

There are three major impact craters on the continent of Antarctica and they are by far larger than any on the other continents. There may be many, many more scientists are unable to detect because of the continent wide glaciation. These three, though highly suspected, and somewhat discovered by satellite probes through the ice, remain unconfirmed.

The smallest is the Bowers Crater. It measures 100 kilometers (62 miles) across and is of unknown origin and unknown age.

The Ross Crater is on Ross Island and is the largest and youngest of the three at 550 kilometers (372 miles) in diameter and dates from the Ogilocene Era, around thirty five million years ago

The oldest is the Wilkes Land Crater. It is on Wilkes Land in the Antarctic and measures 485 kilometers (302 miles) in diameter. It dates from five hundred million years ago.

Biblical Proofs

Since this epigrammatic study has examined the undeniable categorical scientific evidence for profuse prolific catastrophism down through the pre-historic ages, it is time to examine some recorded proofs found in the Bible, many relayed as timely wondrous and astonishing miracles from God. Many people don't realize the many fantastic and incredible miracles recorded in the Bible are quite possibly feasible evidence for the Agent of Mass Destruction. It is time to show this heretofore hidden and shrouded evidence.

The Bible

Starting in *The Book of Genesis;* take a trip down through the Bible and see the Agent in action.

Prior to this, this study has already shown the biblical evidence of the destruction of the Old Earth and the Reconstruction God performed. *Genesis One;* Now this thesis will show how these destructive, indelible and deleterious events kept occurring down through the past historic ages. It is time to show our ancestors faithfully recorded and preserved them for posterity, through a specific religious and mythological filter, one obscuring the unblemished truth of the events when inscribed and printed and feebly interpreted by modern man.

Continual Catastrophes Down Through the Ages

This study has elaborated at some length on the original destruction of the Old Earth, the one requiring the Reconstruction expounded in the ***First Chapter of Genesis***, the one occurring around two hundred and forty million years ago. It is not necessary to go into detail of this event in this chapter also. Here will be discussed the remnants of the original destruction, remnants left as flagellating detritus in the inner Solar System, remnants causing continual scourging trouble for the Earth down through the last six millennia right up to the present. This study will now examine some of these continually plaguing catastrophes.

It must be noted there were more of these incidents of calamitous encounters with the Agent of Mass Destruction in the early millennia of man's written history than there are today, the detritus having dissipated somewhat since those times. Early histories and early mythologies are replete with examples of collisions with wayward remnants of the marauding interloper. This chapter will only discuss the voluminous references to this interloper in the Bible, and a few instances from medieval and modern history. The others, from early histories and mythologies and folklore, will be covered in another study.

From Adam to Moses

Noah's Flood

The next significant event to occur on the Earth of a major catastrophic nature, after the ones previously discussed above, the Agent of Mass Destruction causing almost total destruction of the Earth at the P-T boundary around two hundred and forty million years ago, and after the one at the K-T boundary destroying the dinosaurs and driving them to extinction, was the flood in Noah's day.

In this globally devastating and ruinous event, two significant things happened.

First, the encircling bubble of water, which formed around the atmosphere of the Earth on the second day of the Reconstruction, **Genesis One;** when God ***...separated the waters which were under the firmament from the waters which were above the firmament,*** ruptured. As previously copiously established, the firmament was the atmosphere, and the bubble around the upper atmosphere was burst by a remnant of the Agent, and all the voluminous water came raining back to the surface of the Earth, ... ***the windows of heaven were opened. Genesis 7:11; Genesis 8:2;***

This happened because a remnant of the original collision punched through the envelope of water which was encircling and encompassing the Earth and sent it showering back to the Earth for forty days and forty nights, covering the Earth with water up to the mountaintops.

Second, the subterranean waters, the aquifers and underground seas, were broken up and spewed to the surface adding to the flood of waters. ***...the fountains of the great deep were broken open. Genesis 7:11; Genesis 8:2;***

This happened because of a severe fracturing of the crust of the Earth, which was caused by the collision of a rather large remnant of the original destruction. When it punched a hole in Earth's water bubble, it came crashing down all the way through the mantle into the fountains of the great deep, and the aquifers were spewed out to the surface. It also upset the balance in the oceans and sent their waters cascading over the landforms, adding to the flooding.

Although the Bible tells us the waters came upon the Earth in a mere forty days and forty nights, it took almost a year for the waters to adequately assuage from the land enough for man to once again occupy the beleaguered surface. During this draining of the land, the waters settled back into the ocean basins, filled the lakes, settled in the courses of the rivers, the hydrologic cycle having established itself on the Earth, and returned to the vast underground aquifers, filling them to a greater extent than they were previously.

This provided the necessary water at man's disposal for the millennia since.

It also acted as the method of lubrication at the sub-continental level, which allowed for easier movement of the continental plates in the ensuing continental drift. This caused a faster degree of drift in the earlier stages of the movement than is currently being displayed. With this in mind, it is possible the amount of drift, from the original continent of Pangaea, to the position the continents are found today, represents a much smaller amount of time than scientists are allotting. Scientists think the continental drift was a gradual process being accomplished at a few centimeters (inches) a year since the geologic ages, whereas it may be a process being accomplished in much less time. It is possible with this water as a lubricant, Pangaea was a continent relative late in the geological timeframe. The continents may be drifting a few centimeters (inches) a year today. It may have been faster in yesteryear, maybe meters (feet) or possibly even kilometers (miles) per year.

Time Frames

When reading about Noah's Flood, many people fail to realize one thousand, six hundred and seventy six (1,676) years passed between Adam's expulsion from the Garden of Eden and the Flood in the days of Noah. The Bible covers this whole period in just four chapters, **Genesis Three; to Genesis Six;** with hardly a mention of any major consequential events having occurred during this far reaching time. This was probably because there were no real major events, especially of the catastrophic nature, during this entire time and there were probably no real noteworthy deeds of men, of a moral or righteous nature, to mention. To get an ameliorative appreciation for the length of time this concerns, compare to history and include all the history from *CE 340 to CE 2016*; this would start when Rome was still a world power until the United States of America became a world power. It would include the downfall of the Roman Empire, to Emperor Charlemagne, to Genghis Khan, to the Crusades, to the Reformation, to the Industrial Revolution, to both World Wars, to man walking on the Moon, with all the numerous and divergent inventions, and all of man's scientific and artistic achievements. This was a time of great social, religious and political change, a time of monumental strides forward in man's unending quest for supremacy and a better life.

This is the same amount of time commensurate with this first seventeen centuries of man's habitation of the Earth. And it must be understood, men, in these first seventeen centuries, were just as capable of great and mighty works as the men in these last seventeen centuries.

There is no mention of any monumental towering achievements in these four chapters of the Bible. It credits Cain's descendants with inventions, toolmaking, trades, arts and crafts, but does not delineate or expound on any of these successes or attainments. The only noteworthy behavior seeming to stand out to the chroniclers was the giants in the Earth, and how men contemplated evil with every thought.

It should be noted, again, men lived longer back then, the oldest man, a man named Methuselah reached the ripe old age of nine hundred sixty nine years, and the youngest man mentioned, a man named Enoch was a mere three hundred sixty years old when, it says, "*...God took him...*". ***Genesis 5:24;*** This is a span of over six hundred years for the ages of man on the Earth at the earliest ages. Most of the men mentioned in the Bible during this time averaged an astonishing seven to eight hundred years of age.

The reason for their longevity was the bubble of water which surrounded the Earth, protecting them from the harmful rays of the Sun. When the bubble was burst, during Noah's flood, the harmful alpha and beta rays of the Sun began to affect the ageing process in man and the length of man's days drastically reduced from the seven to eight hundred years range to the one hundred fifty to one hundred seventy five years range in about three generations. Then it gradually reduced further to the one hundred to one hundred twenty years range by the time of Moses. Today, the range of lifespan is about seventy five to eighty years. God said man's lifespan range would be lowered to one hundred twenty years. ***Genesis 6:3;*** He restricted it further in David's day when He stated the life span guaranteed man would then be a mere seventy years. ***Psalm 90:10;*** This is over a ninety percent (90%) reduction in lifespans in just six thousand years. Perhaps the bubble of water was a boon to man and we could have hoped to keep it. Perhaps the New Earth will once again be enclosed with water.

As scientists begin to discover new planets circling other, far away, stars, it will be interesting to see if they find any encircled by a bubble of water, like our early Reconstruction era Earth was so situated.

Continental Drift

Prior to the flood of Noah's day, the world was a very different place. The ***First Chapter of Genesis;*** tells us all the water was gathered into one place, on one side of the planet, and the land was on the other side. The scientists also agree with this, saying there was a single continent, Pangaea, or Gondwanaland, and a single primeval ocean, Panthalassa. Scientists say, in the theory of continental drift, the single land mass began to break apart and the pieces, the continents, began to move away from each other, over time, leading to the configuration of the Earth as it appears today.

The Bible lends some understanding to this process. Why it occurred? How it occurred?

When the waters were separated and there was a bubble of water around the Earth, perhaps even a few meters thick, maybe even more, there was considerably less water on the planet's surface than there is today. As an adjunct, there was significantly less water in the aquifers also, below the surface of the land. When the flood occurred, all the waters from the bubble (the waters above the firmament, the windows of heaven), the waters from the aquifers (the fountains of the deep), and the waters from the sea beds, were unceremoniously thrown upon the landmass, flooding it over forty days and forty nights and lasting for nearly a year. It took this long for the accumulated water to drain from the land. The places it drained into were the sea beds, the lakebeds and the aquifers. The amount of water which drained into the aquifers after the flood was substantially more than the amount of water present before the flood. This prodigious amount of water caused a grander lubrication of the overlying landmass at the level of the foundation, thereby dislodging the continents, which sit on separate shelves, allowing the shelves to start drifting away from each other, continental drift.

Because of the fracturing of the crust, which allowed the flooding from the fountains of the great deep, the continents were already starting to split apart and beginning continental drift. Up to this time, all the ocean water was on one side of the Earth and the land was on the other, one major continent, as God placed it in the third day of the Reconstruction. **Genesis 1:9-10;** The Bible refers to this event as the dividing of the Earth in the days of Peleg, Eber's son. **Genesis 10:25; I Chronicles 1:19;** Is it possible, with the faster rate of drift early on, the continents could have accomplished their drift in the period from the days of Peleg to today?

The actual continental drift did not start happening for some time after the flood of Noah's day, allowing time for the subsiding waters to fill the internal cisterns and for the lubrication provided by the water to release the binding hold of the continental shelves at the level of the bedrock.

Comparing the time between the days of Noah and the flood, to when Peleg was born and lived, there was another few hundreds of years before the continental drift started. This is in accord with geology, for the indwelling waters in the underground substratum would not immediately set the continental shelves free, but would take some time to accomplish, perhaps even hundreds of years.

The Tower of Babel

After men began to replenish the Earth, and multiply upon the Earth, after the devastating largescale flood in the days of Noah, some of the inhabitants, who all spoke the same language at the time, began to fill the fertile valley between the two great rivers which originally flowed from Eden. ***Genesis 2:14;*** Hiddekel (Tigris) and Euphrates. This area was known as Shinar, Mesopotamia, the Fertile Crescent and the Levant from earliest times.

When the people saw the heightened level of their prosperity, they set out to perform great feats of building to leave an enduring legacy. This culminated in the raising of a city with a soaring tower to reach to the Heavens, The Tower of Babel. ***Genesis 11:4;***

God looked down and saw what they were doing and he was sore displeased and He scattered them across the face of the Earth, and they stopped building the city and the tower and left the area. This scattering was far and wide and it resulted in changing their languages and they could no longer understand each other. ***Genesis 11:8-9;*** This is not as hard to envision as one might imagine. Consider the English language since its advent. Today's English speakers would hardly be able to understand English speakers of a mere five hundred years ago, from before the Shakespearean era.

Obviously some major celestial event occurred to drive the people from this area, from the product of their ambition, the city and the tower.

The Famine which Drove Abraham to Egypt

The next anomalous event in the Biblical narrative and along our early historical timeframe is a famine, a grievous severe famine.

There are numerous famines in the scriptures, many of which were grievous, so many, it is hard to see they may have a mutually related cause, which may be easily discerned.

Scholars have always assumed, maybe as a result of the numerous famines, in the early stages of man's history, in the millennia when man first started cultivating agriculture, the weather was not as stable then as now. This assumption seemed to make sense taking into consideration the numerous famines and plagues occurring in those days. For lack of a better explanation, weird weather patterns seemed to fit.

This thesis proposes a better, more satisfactory and improved explanation. These famines were caused when those parts of the Earth which experienced the famines closely encountered a remnant of the Agent of Mass Destruction.

The first mention of a grievous famine in scripture caused Abraham to leave Canaan and abide in Egypt for a time. **Genesis 12:10;** He carried the sum of his livestock and each member of his household on this sojourn to Egypt and he stayed down there during the entire extent of the famine. His livestock was quite abundant as explained later in the tale when Abraham and Lot needed to split up because both their flocks were too large for the grazing land they were both using. At the time he went to Egypt, during this famine, Lot was a member of his household, along with all his flocks, and Lot and his massive flocks and household went to Egypt as well.

It is obvious the event causing the famine in Canaan was not affecting Egypt. It was extremely localized, not something prevailing weather patterns will do in a local area. These two countries are quite close together, if one weather pattern was affecting the sustaining ability of

Canaan, it would most certainly affect Egypt also, as it did in the days of Joseph, when the sorest famine to ever hit the area affected the whole of the Middle East. Joseph took seven years to gather grain against the seven year famine, in this famine, there was no Joseph in Egypt. Since Egypt was fruitful enough to feed many of the neighboring peoples and their flocks and households, Egypt was certainly flourishing during this time when Canaan was destitute. Whereas the famine in the days of Joseph was widespread, affecting the whole area for years, this famine was more localized, affecting only Canaan. The length of time it was prolonged is not disclosed.

After the famine was finished, Abraham and his household returned to Canaan. Previously, mention was made of the size of Abraham's flocks and Lot's flocks, however, Abraham's household was also very, very large. From his household, he was able to recruit an army of almost four hundred men, large enough to defeat the four kings and their armies, who kidnapped Lot. **Genesis 14:12;** His household army was large enough to carry back the spoils from the war and Abraham gave the spoils to the king of Sodom from whom they were taken in the first place. This was a large household.

The Fire Consumes Abraham's Sacrifice

When Abraham returned from Egypt to Canaan, God made him a promise, his seed would number as the stars of the Heaven. During the time of this propitious promise, something occurred which bespeaks a celestial event. Abraham set up an appropriate sacrifice unto the LORD and stood guard over it, driving the birds away. Then he fell fast asleep and a great darkness befell Abraham, and a revulsive horror came upon him. **Genesis**

15:9-12; After his short and disruptive sleep, the Sun set and an impenetrable darkness descended upon the place, a great smoke was seen and a burning fire swept between the sacrifice, consuming it. ***Genesis 15:17;***

This was probably another remnant of the Agent lashing the land of Canaan, and Abraham presented his sacrifice at the right spot at the right time, allowing him to witness this event.

When he fell asleep, the remnant moved into the sky above where he was, causing a great darkness over the area, probably interacting with the Earth's atmosphere. This was the horror he witnessed, the interaction of the body with the Earth, and the resultant electrical discharge which was most surely to follow. Then later, when the Sun set, the darkness descended upon the place. It goes without saying darkness would follow a sunset, therefore this mention of a darkness must have been something more unusual than just night time. It was a deeper darkness resulting from an encounter with the Agent of Mass Destruction, indicating the body was still overhead at this place, and the previous darkness was not caused by nightfall. As the remnant moved closer to the Earth, the electrical discharges probably increased in intensity causing the smoke and the burning fire, and it probably started shattering the body somewhat, cleaving pieces from it, and it was one of these pieces consuming the sacrifice as it plunged into the Earth. Abraham was fortunate to survive considering his close adjacent proximity to this encounter.

Sodom and Gomorrah

The next major injurious catastrophe the Agent of Mass Destruction visited upon the Earth, and man, was the comprehensive destruction of Sodom and Gomorrah and the cities of the plain. This confrontational spoiling event was one occurring in stages.

First, the inhabitants were struck with a pervasive blindness, as if from a brilliant flash, early on the morning of Lot's escape. *Genesis 19:11;*

Second, the rain of fire and brimstone began later in the day after Lot entered the city of Zoar. *Genesis 19:24-25;*

Third, the smoke of the burning was long lasting, ascending up to great heights so it could be seen far away. *Genesis 19:28;*

When Abraham arose the next morning in the plains of Mamre, a long distance west of the plain, he looked toward Sodom and Gomorrah and saw the profuse smoke arising as the smoke of a burning fiery furnace. *Genesis 19:28;* Abraham was acquainted with many inhabitants of Sodom and Gomorrah, including their kings. After he defeated King Chedorlaomer and the four kings, he restored to the kings of Sodom and Gomorrah all they formerly had lost in the War of the Four Kings against the Five Kings earlier. The king of Sodom was especially grateful and offered Abraham all the spoils of his battle, which Abraham absolutely categorically refused. *Genesis 14:21;*

The overthrow of the cities of the plain fractured the mantle of the Earth to reveal the vast stores of salt underlying the area. The resulting Dead Sea is the most desolate place on Earth to this day. Before this cataclysm, it was a vast unfertile plain, called the Vale of Siddim. *Genesis 14:3, 14:8-10;* It was extremely unfertile: and noxious and full of slime pits (bitumen pits) no one could inhabit the bitter place and after the War of the Four Armies against the Five Armies, the fleeing kings hid in the tar pits. After the devastation of the Agent of Mass Destruction, this Vale of Siddim was the Dead Sea.

Southeast of the Vale of Siddim, a fertile place existed where Sodom and Gomorrah were situated and this is why Lot chose it over any other fertile place in the Fertile Crescent when Abraham gave him the choice, and he separated his vast herds from the herds of Abraham. Abraham stayed in the vast plains of Mamre and Lot departed to the fertile lands surrounding Sodom and Gomorrah.

Before this event, there is no reference to the body of water at the center of this plain being called the Dead Sea, only after this event. The reason it is called the Dead Sea is because no life can live in those briny waters, waters so saline, the salt literally floats on the surface. Today, swimmers visit these waters for their restorative effects and swimmers float in the extremely salt water, unable to descend the depths, being forcibly floated again to the surface because of the salty waters.

It is the largest source of salt on Earth, and it is interesting to note, when Lot's wife looked back, she was turned into a pillar of salt. If the cataclysm were blasting salt from the mantle into the air, anyone remaining in the area, not obliterated in the initial blasts, would soon be completely covered with the salt raining down upon the whole area from the upheaval. Such a person would become a pillar of salt. ***Genesis 19:26;***

After the utter destruction of the plain, and with the resultant radiation, Lot did not feel it was safe to remain in the city of Zoar, the only city of the plain which was not destroyed by the rain if fire and brimstone. He took his two daughters and escaped to the mountains and lived in a cave there. ***Genesis 19:30;*** Lot's two daughters thought the whole Earth was destroyed and they were the only ones left, so they lay with their father to repopulate the earth. ***Genesis 19:31;***

It is apparent to those of us who have grown up in the atomic age, the description of the destruction of the cities of the plain was an event closely resembling a nuclear explosion, and was probably nuclear in nature. It has all the earmarks of a nuclear holocaust, a light blinding the population, the rain of fire and brimstone, and the resultant nuclear cloud, which Abraham observed the morning after.

It destroyed the major cities of Sodom and Gomorrah and Admah and Zeboiim, ***Deuteronomy 29:23;*** and all the cities of the plain, except Zoar. All the unfortunate inhabitants who lived in the cities of the plain were obliterated. Everything was utter desolation, throughout the whole plain, the whole area now comprising the Dead Sea basin, including the plain and all the plants growing upon the plain, which had previously sustained Lot's flocks. ***Genesis 19:24-25;***

It is evident an event of nuclear proportions did indeed occur.

Fire and brimstone from the Heavens was an apt description of a large chunk of rock (meteorite), infused with naphtha (hydrocarbons), alight and falling from the sky, upon the area now known as the Dead Sea. It was perhaps a larger remnant of the same body which just the day before had cleaved a smaller chuck which consumed the sacrifice of Abraham. The major part of the Agent was still lurking in the atmosphere, finally breaking up completely and raining down on the Earth at Sodom and Gomorrah. There are other testimonies of events of cataclysmic proportions happening in other parts of the world, China (the Wei Chronicle), the Americas (the Popul Vuh), and Europe (the Edda), during this same time span.

This thesis proposes this marauding interloper was skirting the upper atmosphere, moving all around the world, between the time of Abraham's famine, to the destruction of Sodom and Gomorrah, causing widespread devastation in its wake, finally culminating in the total devastation at Sodom and Gomorrah.

Fallout

After the destruction of Sodom and Gomorrah, after Abraham woke the next morning to "*... see the cloud of its burning ...*", he went south and dwelt in Gerar, **Genesis 20:1;** because the radiation, which was the result of the nuclear nature of the catastrophe, and which drove Lot from his haven in Zoar to the mountains, was now drifting west and it was not safe to stay in Canaan any longer, until the cloud of radiation passed, and the area recovered.

As he did before in Egypt, he misrepresented his wife to the king as his sister. When his deceptive ruse was found out in Egypt, he was asked to leave Egypt, **Genesis 12:20;** and he returned to Canaan. When his ruse was discovered by Abimelech, the King of Gerar, he was permitted to stay and abide in Gerar. **Genesis 20:15;** This was because King Abimelech knew he could not send Abraham back to the land from whence he came, as the Egyptian Pharaoh had done earlier, since Canaan was now undergoing a bath from a radiation cloud. He allowed Abraham to stay in his country, even though Abraham grazed vast flocks and was a serious drain on Gerar's own limited resources.

The Famine in the Days of Isaac

After the destruction of Sodom and Gomorrah, there is not another event until later, in the days of Isaac, Abraham's son.

It was a famine occurring shortly after Abraham's death, and a lot of time passed between this event and the destruction visited upon the plain of the Dead Sea. ***Genesis 26:1;*** This famine drove Isaac to Gerar, and to the court of King Abimelech, and like his father, Isaac also used the same ruse as his father, stating his wife was his sister. This time King Abimelech was wary, and discovered the deception himself, again allowing Isaac to sojourn in his land afterwards. It appears this second great famine was also caused by a remnant of this Agent of Mass Destruction, thereby explaining why King Abimelech allowed him to sojourn in Gerar, instead of returning north to Canaan.

He remembered the radiation from Abraham's day, and how he helped Isaac's father in those days, and since this was a famine caused by another remnant of the Agent, Abimelech allowed Isaac to sojourn in his land.

The Great Seven Year Famine Foretold by Joseph

Pharaoh dreamed a dream and Joseph, who was then in prison, was able to interpret, and the interpretation was seven years of plentiful abundance followed by seven years of sore famine. ***Genesis 41:54;*** Pharaoh placed Joseph over the whole land of Egypt during the seven years of plenty and he stored up the grain against the seven years of famine. When the seven years of famine finally hit, Egypt became the breadbasket of the rest of the world in those days, allowing some people to come and abide in their land, and partake of their food.

This famine was most probably caused by a large celestial event which changed much of the known world's productivity, caused, most probably, by another large remnant of the Agent of Mass Destruction. Even Egypt, which depended on the Nile for its productivity was disrupted. Therefore it must be understood the regular flooding of the Nile was disrupted for those seven lean years.

This could only occur if something celestial, something of a massive nature, something akin to the body breaking up in the days of Abraham, destroying Sodom and Gomorrah, was affecting the Earth. It was probably disrupting the weather and the rainfall because of a near contact. It was adversely affecting the weather, not allowing the rain to come. If this is true, it heralds another close encounter with a large remnant of the Agent, an encounter causing the worse famine ever recorded in the history books.

This also happened in the days of Elijah, a period of no rain, followed by an encounter, which will be discussed later. Though the encounter in the days of Elijah was bad, as will be shown, this one was twice as bad, since this famine lasted twice as long and over a much wider area.

From Moses to the Wandering

After the famine in Joseph's day, four hundred and thirty years later to be exact, another encounter with another massive remnant of the Agent of Mass Destruction occurred on Earth. This happened during the days of Moses, and of the Exodus, and of the Wandering of the Israelites in the Wilderness.

The whole time, from the ten plagues of Egypt brought about by God through Moses, to the conquest of

Canaan by the Israelites under Joshua, there were numerous events of celestial proportions. During this time, the Earth was in almost constant interaction with a very large remnant of the Agent of Mass Destruction.

It is time to enumerate them here.

The Ten Plagues of Egypt

The plagues of Egypt describe manifestations resulting from persistent contact with a remnant, or perhaps with numerous remnants.

The First Plague

The first plague saw the waters turned to blood. ***Exodus 7:14-25;***

This happened when a remnant of the Agent came into contact with the Earth and caused the life-sustaining waters to turn red and mucilaginous and become undrinkable.

The area of the Middle East is replete with name places reflecting the red nature of the area, (i.e. The Red Sea, Eritrea, Edom, etc.) indicating this remnant caused a wide area to be adversely affected by this red-hued tinting. It is possible two remnants, coming into close proximity in the stratosphere or ionosphere of the Earth, smashed into each other, shattering and spreading some of this falling material thoroughly enough to make it a reddish dust all over the area. If the main component of the remnant was iron, the result was a ferrous oxide ash residue covering the whole area in a red rain, causing many areas of the Middle

East to turn red or appear reddish all along the ground and mixing with the waters. This included all their bodies of water, rivers, lakes, ponds and even seas. This would have made the water undrinkable for all the inhabitants of the land, human and animal. This water was necessarily red and of a thick and sticky consistency from this honey-like ferrous oxide being rained down on them, and mixing thoroughly in the potable waters. It would have looked like blood in Egypt and been thick and viscous with a rosy color in the Levant, the land described a short while later as flowing with milk and honey. Honey in these parts is many times red or golden red in color, depending from which flowers the bees gather their pollen.

This pollution of the waters of Egypt resulted in some of the following plagues, also.

The Second Plague

The second plague witnessed the land overrun with frogs. ***Exodus 8:1-15;***

These frogs, which usually inhabit the rivers, ponds, and marshy areas, left those devastated areas when the waters were turned to blood. It was as life-threateningly toxic to them as it was to man, and to all animals. This caused them to literally overwhelm the land, and wholly infest the cities and more populated places, looking for more adequate food sources, since their usual habitat was now a stinking decimated wasteland. They invaded the cities where they could find additional food. Finding none, they perished in the millions there in the cities, under the feet of men. They literally croaked (pun intended).

The Third Plague

The third plague resulted in the dust turned to infesting lice. *Exodus 8:16-19;*

Two different things were happening here. Lice infested the cities and dust from the desert blew into the Nile area because of this event.

Again, this plague was the result of the saturated polluting of the waters, when lice came to the land and left the water areas, where most lice thrive. The areas away from the city became completely dry and arid from the lack of moisturizing water, and the wild Sirocco wind being generated by the close encounter with this remnant bolide. The Sirocco winds whipped up the dust which blew into the city areas from the surrounding deserts. The lice, which formerly lived off the wild and domesticated animals all along the course of the Nile, were caught up in the dust storm and came to the city by way of the whipping and whisking wind. The lice were searching out another nourishing food source as the animals and men of the cities now provided a good available food source for the parasites. Men were abandoning their former habitats which were no longer able to sustain them and they were heading to the cities to find supporting relief and succor.

The Fourth Plague

The fourth plague unfolded as an inundation of all varieties of flies upon the quickly languishing land. *Exodus 8:20-32;*

The flies also came to the land as a result of the first plague, probably on the same hurricane-like winds which previously brought the lice. Also, the flies were attracted to the desiccated carcasses which were in the land from dead animals, like the frogs, who died from lack of water.

Flies multiply when the conditions exist which can sustain a larger population of flies. When carrion are plentiful they multiply greatly, and vice versa. They can go from egg-laying to full-fledged fly in a matter of days, and they seek out dead and dying organisms as host for their offspring. Egypt, at this time, with all the death happening throughout the land from the poisoned waters, was a perfect breeding ground for maggots.

The Fifth Plague

The fifth plague resulted in the death of much of the livestock. ***Exodus 9:1-7;***

It is possible when the two remnants collided, the dust from the collision settled first onto the land, then a few days later, the bolides, or pebble and rock sized pieces, which stayed suspended in the upper atmosphere, began to rain down on the fields along the Nile Valley where the livestock grazed, killing much of the livestock.

It rained death from the skies on the livestock which further reduced the available food supply.

The Sixth Plague

The sixth plague caused men to be afflicted with boils. ***Exodus 9:8-12;***

It is entirely possible the incessant rain of harmful damaging and ruinous bolides was accompanied by a rain of equally destructive and hurtful dust and ash on Egypt at this time. This dust was of ferrous-oxide also, and would have acted as an irritant to the unprotected skin of man, and could also have carried a bacteria causing the wounding and putrid boils.

Scientists are now beginning to seriously investigate if some hearty bacteria could be transported from one celestial world to another by comets and other travelling heavenly bodies. The Agent of Mass Destruction could easily have been carrying some of these tough stalwart bacteria.

The Seventh Plague

The seventh plague was the result of the same collision, when the upper atmosphere was clogged with the dust and ash of the former collision, causing the formation of violent tempestuous storm clouds and was accompanied by noisome hail, voluble rolling thunder and bright glaring lightning. *Exodus 9:13-35;*
It also caused the air and the ground to run with flaming consuming fire as many of these encounters sprayed the whole land with the naphtha substance which was highly flammable.

It was a blazing remnant leaving a trail of blazing fire in the sky and all along the ground where it collided. It is possible many of the encounters with remnants of this Agent of Mass Destruction were with chunks of rock rich in combustible hydrocarbons. This would account for the fire accompanying many of the manifestations, the incendiary hydrocarbons igniting from the fierceness of the encounter, and from the viable friction of its passage through the Earth's atmosphere, like the heat shield on a space vehicle heats up on reentry.

It is interesting to note the Middle East is the richest area of hydrocarbon reserves in the world. This thesis proposes the hydrocarbon reserves in this area were laid down there by this remnant, when it exploded over the Middle East, raining down liquid naphtha along with the stuff it burned. The liquid naphtha hit the sandy soil and seeped down into the reservoirs where it is being tapped and mined today. If an examination is made of all the areas of the world where major hydrocarbon deposits are located today, or were in the past, like Texas, Alaska, Venezuela, The Ural Mountains, The Caucasus, and, of course, the Middle East, the peripatetic meandering trail of this remnant becomes patently apparent, and the scattering of the released hydrocarbons into the subterranean strata of the Earth at these points gives scientists the signposts to this remnants erratic path.

The Eighth Plague

The eighth plague was a plague of locusts. ***Exodus 10:1-20;***
The locusts came upon Egypt because the areas usually providing their food was destroyed and they of necessity came to the fertile areas to eat. Locusts are a hearty species, able to survive on the most meager of subsistence rations. Even though much of the vegetation on the savannah was destroyed, they were able to continue to exist there for many days before they found it necessary to migrate to new more plentiful feeding grounds.

When locusts invade an area for its vegetation, they pick it clean to the roots, and eat everything in their path. This was a swarm of locusts of immense size, coming into the only fertile area left, the only area not bombarded by the detritus of the collision, and not burned by the igniting hydrocarbons. This exacting plague was especially devastating to the remaining population, because it would have meant the destruction of their only remaining provender.

The Ninth Plague

The ninth plague resulted in the whole land being covered in a thick palpable darkness for three continuous days. *Exodus 10:21-29;*
The collision was finally finished, the dust and ash becoming dense enough to shut out the light from the Sun completely, enshrouding the Earth with the cloud, raining hydrocarbons on the Earth, specifically in those spots previously mentioned, burning everything this burning fiery naphtha touched, causing smoke to add to the already choked dense atmosphere, blotting out any light seeping through the impervious gloom.

It is also possible, at this time, the largest remnant of the Agent came between the Earth and the Sun for a period of three days, staying close to the Earth directly over Egypt, blotting out the enlightening beneficial rays of the Sun completely, to the extent one man could not see his neighbor, and a man could not see his hand in front of his face, just like it did in the days of Abraham, when he was offering his promise sacrifice to the LORD as explained above. This darkness was more intense than that one.

The Tenth Plague

The tenth plague brought about the death of the firstborn. *Exodus 11:1-10; Exodus 12:29-30; Exodus 13:15; Numbers 3:13; Numbers 8:17; Numbers 33:4;*

This was probably a result of the final coup-de-gras collision of the remnant causing the three days of darkness. It caused the death of all the first-born sons in Egypt, but somehow missed Goshen, where the first-born sons of the Israelites were dwelling.

Some have speculated the nature of this plague was genetic in nature, a plague born in the remnant, yet able to distinguish the firstborn sons from all the other offspring of a person in the land of Egypt. Somehow, and it is not explained how, in any sense, avoiding the Israelites altogether because of some distinct marker in their genes. As if this remnant contained some potent genetic detector.

This thesis cannot accept this possibility. Instead this thesis proposes this as the true indicator of the cause of these events, God Almighty, visiting upon Egypt His divine judgment for their stubborn ways.

The Plagues Conclusion

Do not be confused here concerning the real reason for these events now referred to as the Ten Plagues of Egypt. The LORD was accomplishing His enduring purpose. This is merely proposing a possible way the LORD did accomplish His unquestioned and omniscient purpose.

God started by removing their source of drinkable water, then step by step He removed their food sources while infesting them with the most pestilential of small animals and insects. Then He sent outbreaks against the larger animals, and finally against man himself. Then He sent the darkness, to heighten their apprehension and finally He took their highly treasured and loved first-born sons.

The fact the first-born of the Egyptians were killed while the Israelite first-born were spared is one indication the LORD was running the show. The fact the first-born sons were killed and the second born sons were not is further evidence of the designs of the LORD. If the remnant were just killing, surely some other sons and daughters would likewise die. It says only the first-born sons of the Egyptians were killed.

Some may say this thesis also removes God from the equation in its desire to explain all the events of those days as products of the Agent of Mass Destruction. Rest assured it definitely leaves God in the equation, as the author and finisher of the Agent of Mass Destruction and all the upheaval it visited upon the Earth throughout man's history, and even before. It is God doing His will and bringing about His designs. This merely explains how God did it. Which is harder for a doubting mind to comprehend, whether God caused these events through a direct intervention to bring about His purpose or whether He set a heavenly body in motion at the very early stages of the Earth's beginning and used this Agent to accomplish His purposes, all His purposes, from the destruction of the Old Earth when Satan fell, to the Reconstruction, to all the many "miracles" God used to keep the Israelites true to the promise through the millennia?

The Parting of the Red Sea

After the death of the first-born sons in Egypt, the next major event of celestial implications was the parting of the Red Sea. ***Exodus 14:13-31; Numbers 33:8; Nehemiah 9:11;*** It is important to note the enormous importance of this event and its adverse effect on many nations at the time, not just Israel. Egypt spent years recovering from this ruinous event, trying to regain her place as a major world power after being soundly defeated by the Agent of Mass Destruction in the days of the Exodus.

Egypt is the only nation to hold a prominent position in world history to experience a complete disintegration and yet recover altogether and regain world superpower status.

The whole night before the wondrous phenomenal crossing, there was a strong east wind blowing all night long. This was caused when another massive piece of the Agent of Mass Destruction approached the Earth, probably the same piece just previously interacting with the Earth, the piece blocking out the Sun for three days just prior to this day, the same piece which caused the hurricane strength winds which blew the dust into the habitable areas from the desert. This large body, a remnant of the Agent of Mass Destruction, sitting immediately outside the Earth's atmosphere, began to approach closer to the surface, interacting with the Earth, affecting the weather patterns, causing this strong east wind, a mighty blast of wind coming into Egypt from the east. ***Exodus 14:21;*** The people saw this body as a pillar of cloud by day and a pillar of fire by night, which indicates it was ablaze with the burning hydrocarbons as it was parked in the stratosphere above the Earth. This means there was a smoking appearance during the day, from the burning mass, and at night, the fire was showing through the abundant smoke and haze.

As the children of Israel approached the Red Sea the invading body moved with them. When the Red Sea parted, the children of Israel saw the pillar move from in front of them to behind them. ***Exodus 14:19;*** It explains the seas were parted with a blast, ***Exodus 15:8;*** and numerous times when noticing these celestial events, there are references to a blast from Heaven, (i.e., Elijah calling down fire from God to consume his sacrifice, ***I Kings 18:21-39;*** King Sennacherib's army being destroyed, ***II Kings 19:35;*** etc.). This blast happened when Moses stretched forth his rod and the parting of the Red Sea occurred.

The water was piled up as a wall on the right and the left. This could be caused when the remnant came into closer contact with the Earth, pushing downward with a tremendous gravitational force on the waters of the Red Sea at this point, pushing them up as a wall on right and left. The gravitational force exerted by this body, directly above the Red Sea, would exert a tremendous downward push on the waters of the Red Sea. This push would cause the waters to be raised straight up into the air on the right and the left of the exerted force, creating a wall on both sides, in accordance with Newton's Third Law. It describes the water being as a wall on the right and the left, ***Exodus 14:22-29;*** in relation to where the children of Israel were standing at the outset, allowing them to cross unimpeded on dry land. This was possible because the waters of the sea at the place of the crossing were completely dried up and because the pressure would not necessarily have been too great to impede a man or beast from walking yet would still have been of sufficient strength and magnitude to separate the walls of waters because of the surface area exposed to the force. ***Exodus 15:8;*** Along with the great pressure was a tremendous heat associated with the interaction of the body with the Earth. The two combined would have allowed the children of Israel to cross the sea bottom on completely dry soil.

This downward pressure would also suitably explain why the wheels of the Egyptian chariots were dismantled when they attempted to cross at the same spot as the Israelites. It says their chariots were driven down heavily and the wheels came off, and then the walls of water returned and they were drowned. *Exodus 14-24-27;* The Israelites were able to walk across the dry sea bed unhindered because the force was not strong enough to make them unable to move forward, and they were not in wheeled vehicles, *Exodus 12:34;* they were on foot. *Exodus 12:37;* The Egyptian army however were riding chariots and the wheels were unable to operate under the intense downward force, and it says the LORD took off their wheels, dismantled the axle mechanism. *Exodus 14:25*; The LORD was also spoken of as in the pillar of cloud and fire. It can be assumed this was what the narrator meant when he says the LORD took off the wheels, the pillar of cloud and fire took off the wheels, the remnant of the Agent took off the wheels.

When the Egyptians were crossing to follow the Israelites, the body moved away from the Earth and as it moved away, there was an electrical discharge between the Earth and this massive remnant. Whenever two celestial bodies with electrical charges move into close proximity to each other, they exchange charges, and as they move away, they release this attraction with an electrical discharge. This discharge would have caused the waters formerly pushed apart to collapse back into their basin, drowning anyone caught in the midst of the waters. The Israelites were given sufficient time to exit the sea bed on the other side, however, the discharge caught the Egyptian army in the midst of the sea, and they were drowned.

The body stayed in Earth's vicinity and caused many of the events of the Wandering. When the body sparked with the Earth and moved away, the waters would be released from their suspended state, probably by the electrical discharge, and surely because the body was no longer exerting its gravitational forces on the waters of the

Red Sea, and therefore the waters returned and drowned the Egyptian army. It avers unequivocally it was God who destroyed them, if there is any question about the cause of all these events. ***Deuteronomy 11:4;***

    Before the waters returned upon the Egyptians, God looked down on them from out of the pillar of cloud. ***Exodus 14:24;*** As the body pulled away from the Earth and as it was leaving the vicinity of the Earth, it once again came closer to the Earth, actually invading into the atmosphere of the Earth. This caused a tremendous gravitational force to be exerted down on this area of the Red Sea. The Bible explains this by saying God took off the Egyptians chariot wheels. ***Exodus 14:25;*** Then as the Sun broke the horizon the waters returned and drowned the Egyptian army. ***Exodus 14:27;***

    The depths covered them. ***Exodus 15:5; 19; Joshua 24:7; Nehemiah 9:11;***

    And this description overrules all the naysayers who state the place of the crossing was the Sea of Reeds, a shallow sea north of the Red Sea. This study posits it was the Red Sea, a place with depths. There was also a blast associated with the returning of the waters, ***Exodus 15:8;*** and also a wind was blowing when the waters returned. ***Exodus 15:10;*** There was also an earthquake in the midst of the sea, where the Egyptian army was, because it says the Earth swallowed them. ***Exodus 15:12;*** This could also be a result of the body being pushed closer to this area just before it separated from the Earth. The Bible says the Egyptian army sank in mighty waters like a stone. ***Exodus 15:5-10; Nehemiah 9:11;***

The Pillar of Cloud and Fire

The remnant remaining in Earth's vicinity then became what the Israelites saw as a pillar of cloud by day, and a pillar of fire by night. This was probably a result of it being in an elongated shape after the encounter with the Earth at the crossing of the Red Sea. God used this remnant to lead the children of Israel throughout the wandering in the wilderness. It affirms it went before them day and night. ***Exodus 13:21-22; 33:9-10; 40:38; Numbers 9:15-23; 14:14; Deuteronomy 1:33; 31:15; and Nehemiah 9:12; 9:19;***

When the cloud moved, the Israelites moved. ***Numbers 9:15-22; 10:11-12; 10:34;*** Sometimes this remnant remained in one place at the upper reaches of the atmosphere, and sometimes it would move along, always staying in the vicinity of the Fertile Crescent, and the land of the wandering.

Throughout the wanderings, the people experienced the hand of the LORD about them, and many of the descriptions can be related to the celestial remnants.

It says, in many places, God delivered Israel by ***"...trials, signs, wonders, war, a mighty hand, an outstretched arm, and great terrors, and great judgements..." Deuteronomy 4:34; 5:15; 6:21-22; 7:8; 7:19; 11:2; 29:3; Exodus 3:19-20; 4:9,21; 4:30; 6:1; 6:6; 7:4-5; 13:3; 13:9; 13:14; 13:16; 15:6; 15:12; 15:16; Numbers 14:11; 14:22; Joshua 24:17; II Kings 17:36; I Chronicles 16:12; 16:14; 16:24; Nehemiah 9:10; 9:17;***

Israel was delivered by *trials* because ordinary life was crushingly hard in this barren wilderness, not just in the search for daily sustenance, but also from the continual effects of this Agent of Mass Destruction.

Signs and wonders because they definitely saw many signs and wonders in their meandering travels. One can think of no other people who were ever delivered from such lengthy and insufferable bondage in such a miraculous and mighty way?

War because there were numerous hostile neighbors (i.e. the Amalakites) **Numbers 14:42-45;** also on the move at this time, searching for a better homeland since this encounter with the Agent of Mass Destruction made their old homeland of little use, entirely burned and fully destitute. The children of Israel went to war with many of these other banished nomadic tribes, both in their wandering, and at the end of their journey to take the Promised Land.

A mighty hand because the LORD was effectively delivering the people, not themselves, not another people, but the LORD Himself. Never were so many people displaced from so many areas by one event before in man's lugubrious history, and God was doing this to accomplish His purposes, bringing the promise He made to Abraham to fruition, bringing the children of Israel into their assured inheritance.

An outstretched arm because the children of Israel were seeing the remnant as an elongated body, and there were many times in their wanderings when it looked to them like an arm, stretched across the sky, they saw it as a deliverance arm, there to smite their enemies.

Great terrors because the events the children of Israel were witnessing in their hurried flight and in their not at all hurried wandering were terrible and awe inspiring indeed, and the miracles were accompanied by frightening portends in the sky. The very Earth itself seemed to be absolutely disrupted, and this encounter lasted over fifty two years, which caused the people to weary if it would ever end.

Great judgments because they saw the mighty hand of God deliver one nation after the other into their hand,

and they realized it was God's judgment against the wicked inhabitants of the lands through which they passed.

These scriptures define an encounter with celestial objects, and this occurred throughout the wandering. This is evidence of the Agent of Mass Destruction

Mount Sinai

It definitely appears the children of Israel, and Moses especially, witnessed a very close encounter with this remnant from the Agent of Mass Destruction. This happened at the Mount of the Lawgiving, Mount Sinai. The descriptions of the time they spent at Sinai could be nothing else. They describe the classic signs of the encounter.

The descriptions include: thunders, ***Exodus 19:16; 20:18;*** lightning, ***Exodus 19:16; 20:18;*** thick clouds on the mountain, ***Exodus 19:16; 24:15-16; 24:18; 34:5; Deuteronomy 4:11; I Kings 8:12; II Chronicles 6:1;*** the voice as of a trumpet exceeding loud, ***Exodus 19:16; 20:18;*** the mountain smoked, ***Exodus 19:18; 20:18;*** the presence of a tremendous amount of fire, ***Exodus 19:18;*** earthquakes, ***Exodus 19:18;*** the sound of the trumpet for a long time growing louder and louder, ***Exodus 19:13; 19:16; 19:19; Deuteronomy 4:11; 9:15; 10:4;*** thick darkness on Mt. Sinai, ***Exodus 20:21; Deuteronomy 4:11; 5:22;*** the sound as of a voice was heard out of Heaven, ***Exodus 20:22; Deuteronomy 4:36;*** and a great fire on Earth, ***Deuteronomy 4:11; 4:36;*** and there was the sound of a great voice out of the fire. ***Deuteronomy 4:12; 4:15; 4:33; 4:36;*** These are classic descriptions of an encounter with a celestial object, causing havoc on the Earth.

At the time of the lawgiving, God was using the remnant to envelope Mount Sinai with all these elements at the time Moses ascended the mountain, to set Moses apart from the rest as the leader of the Israelites. The remnant of the Agent descended upon Mount Sinai and caused the thunders, the lightning, the thick cloud on the mountaintop and the smoking mountain, resulting in thick darkness all around the mountain and out into the expansive plain surrounding it. This smoke was probably volcanic in nature as earthquakes wracked the mountain from the encounter and fire spewed out of the cracks onto the mountainsides and possibly out into the plain as well. Along with the fire running along the ground, again reminding us of the hydrocarbon nature of the encounter, there were loud voices heard in the roaring fire, a possible earthquake as the fire ran along fissures, igniting naphtha sources, which previously seeped deep underground on an earlier encounter, with loud sounds like mighty voices. As the remnant remained in close proximity to the mountain and to the Earth, the stresses caused the earthquakes and would be accompanied by extreme noises, like the sound of trumpet blasts, or voices out of the sky and from the distressed ground.

The Destruction of Dathan and Korah

There arose a rebellion in the camp of the children of Israel, and Dathan and Korah questioned the autocratic leadership of Moses. As a result, Moses called all the people together to allow the LORD to choose who would be his spokesman. Dathan and Korah and the rebels did not assent to come up to the gathering, so Moses called for all those among their rebellious camp who wanted to serve the LORD, to separate themselves from the camp of the recalcitrant dissidents.

Then the cruel calamity fell. The Earth swallowed the rebels and all their disobedient households and their plentiful belongings. This was a massive earthquake, again caused by our remnant orbiting right above Mount Sinai, exerting tremendous stress at this point on the plain, swallowing up a host of the children of Israel who stayed away from the protective confines of the mountain where Moses called them. **Numbers 16:30-33; 26:9-10; Deuteronomy 11:6;**

Right after the earthquake, an incinerating fire came forth from the LORD and consumed two hundred and fifty men offering unholy incense again showing the volatile incendiary nature of the naphtha flowing in the area from the interloper in the sky. The spark used to ignite the offered incense also ignited the naphtha in the air and running along the ground, consuming the two hundred and fifty men offering the incense. **Numbers 16:35; 26:10;** It says many of the children of Korah did not die in this calamity. They must have removed themselves from the camp of Korah and joined Moses on the foothills of the Mount. **Numbers 26:11;**

The Wandering of Israel in the Wilderness

Along with the Israelites wandering from their place in Goshen, Egypt, there were many tribes and peoples wandering around for a new place to inhabit since their homeland had been destroyed. An example is the Amalakites whom the Israelites militarily encountered, in the wilderness as discussed above. **Exodus 17:8;**

From The Wandering to the Conquest

The Land of Milk and Honey

Another proof showing the lands of many peoples were being disrupted at this time in history is the Levant itself. When Moses sent the twelve spies up to the land to spy out the inhabitants, ***Numbers 13:1-33; and Deuteronomy 1:24;*** the report they returned was of a land flowing with milk and honey. ***Exodus 3:8; 3:17; 13:5; 33:3; Leviticus 20:24; Numbers 13:27; 14:8; 16:14; Deuteronomy 6:3; 11:9; 31:20; and Joshua 5:6;***

Egypt, the land from which they just departed, was also described as flowing with milk and honey. A thick porous fluid of milky consistency was flowing in the land, in the rivers and bodies of water, presumably after the blood water had flowed out from the days of the plagues. ***Numbers 16:13;*** This was caused from the encounter of the Earth with this remnant, a remnant with vast hydrocarbon deposits. The hydrocarbons from this mini-world and the oxygen in our atmosphere were mingling, causing a rain of fire on the Earth (hydrocarbons) in many cases, and a rain of a thick milky like substance (carbohydrates) in others.

If one examines the mixture of hydrogen and carbon in different combinations, it reveals there are two different manifestations from the combination of hydrogen and carbon.

One manifestation is hydrocarbons which are the composition of all of the fossil fuels used in automobiles today and was the source of the substance known as naphtha back then.

The other is carbohydrates which are edible as breads and foodstuff, such as the unfamiliar manna.

There are numerous instances where the chemical combination of hydrogen and carbon resulted in hydrocarbons, the fuels, the naphtha, lighting fires all along the ground, and the current deposits of fossil fuels in the ground is a result of this specific combination of chemicals from the Agent of Mass Destruction, not necessarily the decomposed bodies of the dinosaurs.

Another combination of hydrogen and carbon from these encounters is a rain of carbohydrates, another form of hydrogen and carbon mixing. These hydrogen carbon combinations (carbohydrates) were manifested in both the manna the children of Israel gathered every morning, ***Exodus 16:35; Deuteronomy 8:3; 8:16; and Nehemiah 9:15; 9:20;*** and in the appearance of the land, flowing with milk and honey. If the hydrocarbons were falling on the rivers and ponds, the appearance of those water sources would be as milk and honey. It appears the mixture of these carbohydrates and the water of the land was a potent combination, which affected the size of the crops grown in these areas. ***Numbers 13:23-27;*** an***d Deuteronomy 1:25;*** The falling of the miraculous manna ceased on the day after Passover, shortly after they crossed over the Jordan, and began the Conquest of Canaan. ***Joshua 5:12;*** The reason for this sudden stoppage of the rain of manna was because the remnant left the orbit of Earth at this time, heading back to space where it was yet to wreak more havoc.

When the spies came up to spy out the land, they returned saying the land flowed with milk and honey and they brought back large luscious fruit and vegetables to emphasize their point. They also reported the land was full of enormous and frightening giants and the Israelites did not stand a chance against them, only Joshua and Caleb believing the whole land could be theirs, according to God's explicit word. Then a noteworthy event occurred. It says the anger of the LORD was kindled against the spies ***Numbers 32:10;*** and ten of the spies died of plague.

Numbers 14:37; Plague was another common manifestations of an encounter with this remnant.

This news caused the people to despair and they lost faith in the LORD, therefore the LORD caused them to wander for forty years, until this generation of skeptical unbelievers perished. God was saying he didn't want this unbelieving generation to inherit the Promised Land because of their disobedience. It appears the LORD may have also harbored another reason to cause the people to wander for forty years. It was during these forty arduous years when many more encounters with the remnant of the Agent of Mass Destruction were causing destruction throughout the area of the Levant, the area formerly described as flowing with milk and honey and filled with giants.

By the end of the forty years, when the children of Israel finally took the land, it was no longer described as a land flowing with milk and honey, the encounter was finally over, and the carbohydrates were no longer pervasively falling. Also, there were not many threatening giants left in the land, as opposed to the land crawling with them before the wanderings. During these forty years, the majority of the giants, and the other inhabitants of the land, were killed, or driven out by the encounters with this remnant, setting the stage for the Israelite occupation of the land. Those who remained were few, and they lived mostly in walled cities. Holy Writ affirms the Angel of the LORD drove out the inhabitants of the land. ***Exodus 33:2;*** The Bible also lists the inhabitants who were driven out as the Canaanites, the Amorites, the Hittites, the Perizzites, the Hivites and the Jebusites, many of whom were of the giants. ***Nehemiah 9:24;***

These inhabitants were driven out by the onslaught of this remnant, the terrible judgments of the LORD wrought by this remnant. When the people of Israel came in to conquer the land after their wandering in the wilderness, some of these peoples were again in the Promised Land. Having found nowhere else to settle during this time, they returned to their previous land.

More Wandering

When the children of Israel moved around Seir and Paran *Numbers 10:12*; it says the Earth trembled; *Judges 5:4; and Psalms 68:8*; it poured rain *Judges 5:4; and Psalms 68:8;* and the mountains melted. *Judges 5:5; and Psalms 68:8;* This was another reference to the celestial event, causing earthquakes and mountains to melt, and a pouring rain.

Plagues

Shortly thereafter, the Bible confirms an angel swept the camp with plague, another reference to a deadly plague associated with this marauder, this Agent of Mass Destruction. *Exodus 32:34-35*;

After the LORD supplied the people with quail in the camp, an event occurred caused by a mighty wind in the evening, *Exodus 16:12-13; and Numbers 11:31-32;* the LORD struck the people with another slaughtering plague. *Numbers 11:33;*

There was enough quail brought up to the camp to feed at least six hundred thousand men, women and children. *Numbers 11:21*;

The LORD, on another mentioned occasion, again struck the people with widespread plague whereupon fourteen thousand and seven hundred died of the plague. *Numbers 16:46-50;*

These plagues were possibly all caused by encounters with the remnant because at every close encounter, there follows a plague.

Fires

In another scripture it refers to a fire from God taking the offering ***Leviticus 9:24*** and shortly thereafter a fire killed Nadab & Abihu when they offered incense. ***Leviticus 10:2; 10:6; and Numbers 3:4; 6:2; 6:61***;

Also, a fire manifestation from this encounter caused the consuming of many on the extreme outskirts of the camp, indicating there were fires burning in the areas of the plain behind the Israelite camp. ***Numbers 11:1-3***;

On another occasion, the people were being killed with fiery serpents, ***Numbers 21:6-9; Deuteronomy 8:15;*** only to find relief when Moses made the fiery bronze serpent at God's direction. ***Numbers 21:8-9;*** This bronze serpent later became an abhorrent idol to the Israelites, called the Nehushtan, and was finally destroyed by King Hezekiah. ***II Kings 18:4***;

The Host of Heaven

God commanded the Israelites, during the lengthy wandering in the wilderness, not to worship the Host of Heaven. This was something in which they would not be very successful during their long sinful and rebellious history. ***Deuteronomy 4:19; 17:3; 2 Kings 21:3; Romans 1:25;***

It is clear in scripture the Sun, Moon and stars are what is meant by the Host of Heaven, not the angelic host. ***II Kings 23:5;*** Instead, God emphasizes it is the Host of Heaven which worship the LORD. ***Nehemiah 9***:6; God made a special commandment against the worship of the Host of Heaven because He wanted to assure man would worship Him, and Him alone, for the great monumental events which were then poignantly shaping the history of the children of Israel. He wanted to be sure anyone who could see the Agent of Mass Destruction and its remnants, and realize this Agent was causing much of the trouble, would not be tempted to worship the Host of Heaven.

And the warning comes down through the ages to man today. Sit and sift through the evidence and come to the conclusion there was an Agent of Mass Destruction, but worship the LORD, and not the Host of Heaven.

From the Conquest of Canaan to the Judges

The Parting of the Jordan River

Just as the parting of the Red Sea was an event brought about by the remnant of the Agent of Mass Destruction, the parting of the Jordan River was likewise. ***Joshua 3:13-17;*** The waters did not flow downstream for an extended period of time, long enough for all the people of Israel to completely cross. Some have conjectured a slice of the bank upstream broke off and blocked the river because this happened in the *CE 1930's* to give a plausible precedent for this possible conjecture. However, the scriptural narrative does not even slightly enforce this conjectured view.

The scripture tell us the waters were cut off from above and stood as a heap way back upstream. ***Joshua 3:13; 3:18;*** The Jordan River was in flood stage at the time, ***Joshua 3:15; 4:18; I Chronicles 12:15;*** further discounting the slice of the bank theory.

What many people don't realize is the Jordan River was witnessed as being parted this way three times, not once. It was parted this time for the people of Israel, an event with long duration for all the people to cross. The other two times were of much shorter duration and both of them were probably caused by a single encounter with the remnant of the Agent.

The other two times were in the days of Elijah and Elisha. Once when Elijah and Elisha crossed east to go out to the wilderness, where Elijah was to be taken up to Heaven by the chariot of Israel, and then again when Elisha was returning west after the fateful encounter. A look at these two encounters will be undertaken in another part of this narrative. All three partings were accompanied by many spectacular celestial events. ***II Kings 2:8, 14;***

The Conquering of the Land of Canaan

There were many celestial events occurring when the children of Israel were conquering the Promised Land.

The Fire in Heshbon

The first event occurred as the children approached the Promised Land on the far west bank of the Jordan River, before they crossed over into the Promised Land. It describes how a blazing fiery conflagration came out of Heshbon, the city of King Sihon. ***Numbers 21:28; Deuteronomy 2:30-35; Nehemiah 9:; and Jeremiah 48:45;*** This fire totally consumed many separate places, and destroyed many indigenous people, the inhabitants who were allying themselves against the invading Israelites. It lists the following places destroyed by this fire: Heshbon, Ar of Moab, ***Numbers 21:28; Isaiah 15:1;*** it laid waste as far as Dibon and Nophah, *Numbers 21:30* and it consumed the near corner of Moab. *Jeremiah 48:45;* It also says there were other notable places destroyed in this same fashion. King Og of Bashan was destroyed the same way as Heshbon. *Numbers 21:33-35; Deuteronomy 3:1-6; and Nehemiah 9:22;* King Balak and Moab were sore afraid of Israel, dreading their approach. *Numbers 22:3;* They proclaimed lamentably, *"Israel shall consume the land," Numbers 22:4;* and they rightly attributed the consumption of the entire land to the children of Israel and their mighty God.

Later, when King Balak hired Balaam to prophesy against Israel, Balaam also encountered a remnant from this destroying marauder. He saw an angel of the LORD with a drawn sword, a manifestation of the remnant with the appearance of a drawn sword. *Numbers 22:22-35;* Balaam asked *"Who shall live when God does this?" Numbers 24:23;* Then God's fierce anger was kindled against Israel, itself, and twenty four thousand died of a ghastly plague. *Numbers 25:3-9; 25:18; 26:1; 31:16;* It declares, many times in scripture, God is an all-consuming fire. *Exodus 24:17; Deuteronomy 4:24; 9:3*;

The Fall of Jericho

When the Israelites crossed over the Jordan River, they began their conquest of the entire land of Canaan at Jericho. As they approached Jericho, Joshua saw the Army of the LORD, which was probably some manifestation of the remnant. *Joshua 5:13-15;* The Fall of Jericho was likewise an event accompanied by celestial events. The general consensus holds, during the fall of Jericho, there was a great earthquake. *Joshua 6:5-21;*

The Bible describes the city walls falling at the long sound of the ceremonial ram's horns, the melodious sound of the trumpet and the ear-shattering shout of the multitudinous people. ***Joshua 6:5;*** The explicit implication is there was a tremendous earthquake accompanying all the tremendous noise the Israelites were making at the behest of Joshua, God's spokesman, and their new leader. This wall destroying earthquake was accompanied by the loud, long sound of a trumpet blast. This was a groaning of the land created by the stresses exerted by the remnant of the Agent of Mass Destruction.

The Sun and the Moon Standing Still

Then after they departed Jericho, they encountered the Amorites, who came out to meet the Israelites and engage in battle. The LORD fought with the Israelites and the Amorites were slaughtered over a large area of the Levant, from Beth Horon to Azekah. It says great stones fell from Heaven, and more Amorites died by these stones than by the sword. ***Joshua 10:11;***

Then comes the seminal event most definitely proving this theses hypothesis, the remnant was affecting the Earth in major and catastrophic ways.

In the valley of Ajalon, as the tiresome battle had proceeded all day, the Sun began to set while the Israelites were on the advantage. Joshua wanted to pursue the enemy through to victory, but with the regular diurnal setting of the Sun, he would have to stop, which would give his enemy time to refresh, rearm and regroup. He ordered the Sun and the Moon to stand still while they finished the ensuing battle, and the difficult enemy, the Amorites. ***Joshua 10:12-13;***

It confirms the Sun and Moon stood still in the midst of Heaven for about a whole day, an entire twenty four hour period. ***Joshua 10:13;***

In order for the Sun and the Moon to stand completely still in the sky, it was necessary for the Earth to stop regularly rotating on its axis. The only thing able to effect the Earth's quotidian rotation on its axis would be if it encountered a large enough force to affect such preciseness of rotation. This was in all logical probability exactly what happened, another major encounter of the Earth with the larger remnant, an encounter of sufficient force to stop the axial rotation of the Earth for about a whole day.

Catastrophes in the Days of the Judges

There were many significant consequential events of celestial proportions in the days when the Judges ruled in Israel. In the days of Deborah and Barak, when they defeated King Sisera, ***Judges 4:7-22;*** it declares the stars fought from their course. ***Judges 5:20;*** It says the ancient torrent of Kishon swept them away. ***Judges 5:21;*** This is a description of the remnant again affecting the Earth.

In the days of Gideon, it states his sacrifice was consumed by fire from the rock. ***Judges 6:21;*** This was probably a result of another fire from the naphtha previously deposited within Earths porous rocks by this remnant.

From Judges to Solomon

The Camp of the Israelites

When the Israelites were prepared to bring the sacred Ark of the Covenant to their new capital at Jerusalem, there was no house for it there, since the temple was still a generation in the future. King David wanted to build the house, but God told him the task would be for King Solomon, his son and heir.

The Israelites were noted for bringing the Ark out in front of their lines during battles, convinced if the Ark was present at the front, God would do the fighting for them. The Ark was in their camp when they went out to battle. *I Samuel 4:5-7;* While the Ark was in the camp, the Earth under their feet shook, *I Samuel 4:5;* and it says God was present in the camp. *I Samuel 4:7;*

This was how the Israelites lost the Ark, because they had it with them whenever they went forth to battle an enemy. The Philistines captured the Ark after one of those calamitous battles. *I Samuel 4:11-22;* While the Philistines possessed the Ark, the remnant of the Agent of Mass Destruction visited them in a disastrous and odiously heinous way. Five of the Philistine cities were affected by this celestial visitation while the Philistines held the Ark, Ashdod, Gaza, Ashkelon, Gath and Ekron. *I Samuel 6:17;*

A statue of their main deity, Dagon, was overthrown every morning. *I Samuel 5:3-4;* It explains the hand of the LORD, (the remnant) was heavy upon Ashdod. *I Samuel 5:6;* A dreadful plague came upon the city, and men broke out in noisome putrid tumors. *I Samuel 5:6*;

It declares the hand of the LORD was heavy against Gath, where the terrifying and forbidding giant, Goliath, was born. It again relates the plague presented with many putrescent tumors and also there was a great terrifying destruction. *I Samuel 5:9;*

It then enumerates a great destruction in Ekron, with the accompanying plague of pustules and tumors. *I*

Samuel 5:11-12; This was probably bubonic plague, maybe even the first documented case of bubonic plague, because the Bible plainly says the plague was in the land because of many invading rats. ***I Samuel 6:4-5;***

It avers there was a great plague in Beth Shemesh because the people looked into the Ark, and fifty thousand and seventy people died. ***I Samuel 6:19***; Later, the Philistines put two and two together and realized it was because of the Ark this evil was being visited upon them, and they arranged to return the Ark to Israel.

When Jonathan and his armor-bearer were defeating the Philistine garrison, ***I Samuel 13:3; 14:6-15;*** there was a great earthquake, ***I Samuel 14:15;*** caused by the remnant. It was of great strength and over such a wide area the Israelites saw the danger, ***I Samuel 13:6;*** and hid themselves in caves, thickets, rocks, holes and pits. ***I Samuel 13:6;*** Some crossed over the Jordan to hide from the trembling Earth in the desert wilderness ***I Samuel 13:7;*** the Earth trembling for a long time. ***I Samuel 13:7;*** It confirms the noise of the earthquake increased ***I Samuel 14:19;*** and there was great confusion ***I Samuel 14:20;*** and men scattered. ***I Samuel 14:16;***

It was then every man's sword was against his neighbor, because of the confusion the manifestation of these catastrophes brought. ***I Samuel 14:20;*** It explains how King Saul's spirit was greatly troubled by the LORD. ***I Samuel 16:14-23; 18:10; 19:9;*** Because of this troubled spirit, King Saul went to see the forbidden and closeted witch of Endor, ***I Samuel 28:7-20; I Chronicles 10:13;*** King Saul compelled her to conjure up the deceased spirit of Samuel for some much needed answers. ***I Samuel 28:12-15;*** As the witch of Endor was conjuring up the spirit of Samuel, it confirms she saw 'elohim'. ***I Samuel 28:13***; It is very possible she was given a glimpse at the Agent of Mass Destruction.

King David and the Remnant of the Agent of Mass Destruction

When King David defeated the Philistines at Pasdammim, *I Chronicles 11:13;* it was described as a great deliverance from the LORD. *I Chronicles 11:14;* It is highly likely this sweeping victory was accompanied by an encounter with the remnant.

Again when King David defeated the Philistines at Baal Perazim, *II Samuel 5:19-25;* it explains the LORD struck them down, *II Samuel 5:24;* referring to the pernicious influence of the remnant. It says they were defeated by an overflowing flood of waters, *II Samuel 5:20;* and as King David lay under a grove of mulberry trees, he heard the clattering sound of armies marching at the tops of the mulberry trees, in the sky directly overhead. *II Samuel 5:24; I Chronicles 14:15;* The noise made by the encounter with the remnant was as the sound of a marching army. It brought about a flood of waters to help in the Israelite victory.

Again, in another instance, the Bible refers to the LORD showing His displeasure against the people of Israel. It says the LORD broke out against the people. *I Chronicles 15:13;* The implication is the people were visited with a plague, and it has already been conjectured the connection between plagues and the remnant is evidence of the Agent of Mass Destruction.

Famine in the Days of King David

There was also a mighty famine in the days of King David equal to the widespread famines in the days of the patriarchs, Abraham, Isaac and Joseph. ***II Samuel 21:1;*** This famine, like those in olden days, was a result of the encounter with the remnant.

King David Sees the Angel over Jerusalem

King David, at one point, wanted to see how many men he commanded in his army. To find this out he prepared to conduct a census. The LORD admonished him not to do this, but King David did it anyway. ***II Samuel 24:1-10; I Chronicles 21:1-4; 27:24;*** As a punishment for this sin, God gave King David three choices of punishment on the people of Israel. ***II Samuel 24:13; I Chronicles 21:10;*** He could have seven years of famine ***II Samuel 24:13; I Chronicles 21:12;*** which mirrored the days of Joseph and the terrible famine bringing Israel into slavery under Pharaoh for four hundred and thirty years, or he could choose three months of being slaughtered by their enemies, ***II Samuel 24:13; I Chronicles 21:12;*** which is something Israel never suffered in their short history and probably would not survive, or at least would be sorely pressed to recover from after such a demoralizing event. The third choice was three days of plague, ***II Samuel 24:13; I Chronicles 21:12-17;*** which is the exact incubation period for the bubonic plague. King David chose the three days of plague, ***II Samuel 24:15-16; 24:25; I Chronicles 21:13; 21:22;*** and when the plague started, King David saw the angel of the LORD over Jerusalem with a drawn sword. ***II Samuel 24:17; I Chronicles 21:16; 21:20; 21:27; 21:30;*** It informs us the sword of the LORD was over Israel, ***II Samuel 24:17; I Chronicles 21:12-16;*** and the angel of the LORD was destroying in Jerusalem during the three days of plague. ***II Samuel 24:16; I Chronicles 21:12; 21:15;*** Seventy thousand died of this plague. ***II Samuel 24:15; I Chronicles 21:14;*** This was caused by the remnant, as were many of the other plagues in scripture. This remnant was laden with disease (bacteria and viruses) and death for the inhabitants of the areas where it made contact.

King Solomon and the Temple

When King Solomon built the temple for the House of the LORD, there was another persistent manifestation of the remnant upon Israel. It explains Ornan saw the Angel of the LORD, and his four sons hid from the terrible sight in abject fear. ***I Chronicles 21:20;*** It further avers when the temple was completed and they offered the first sacrifice, a consuming fire from the LORD fell on the offering. ***I Chronicles 21:26; II Chronicles 7:1-3;*** This was an obvious manifestation of the remnant raining fire down on Earth, as it did to a greater degree over Sodom and Gomorrah, and more persistently in the days of the Wandering of the children of Israel in the wilderness. It also stated when King Solomon built the temple, ***I Kings 6:1; I Chronicles 22:16; 28:6; 28:10; 28:19; II Chronicles 2:1; 2:6-9; 2:12; 3:1-2; 6:9-10;*** a cloud descended and covered the temple ***I Kings 8:10-12; II Chronicles 5:13-14;*** and the glory of the LORD filled the temple, ***I Kings 8:12; II Chronicles 5:14; 7:1-3;*** which could also be a reference to effects of the remnant.

From Solomon to Elisha

The Kings of Israel

There was another noteworthy incident in the days of King Jeroboam. He sacrificed on an altar he ordered built at Bethel and the LORD miraculously withered his hand, then the altar was split asunder and choking ashes poured out from the split in the altar. Then King Jeroboam's hand was astonishingly restored. *I Kings 13:4-6;* This could definitely be a result of the remnant crashing down onto the altar King Jeroboam built. Some sort of plague also hit at this time, withering King Jeroboam's hand.

It explains how King Jeroboam provoked the LORD to anger, *I Kings 14:9;* and God brought disaster on the ten northern tribes, the LORD cut off every male from King Jeroboam. *I Kings 14:10;* This resembles the tenth plague of Egypt, where the first born son was taken in this lethal plague. It further confirms the feral dogs ate the dead in the city, and the birds ate the dead in the fields, *I Kings 14:11;* indicating there were dead all over the land, a truly devastating plague.

It also says in the war between King Abijam and King Jeroboam, *I Kings 15:6; II Chronicles 13:3-20;* the LORD struck Israel before the city of Jerusalem. *II Chronicles 13:15;* The implication of these scriptures is the LORD caused another devastation in Israel, another visitation of the remnant.

In another war between Ethiopia and King Asa, it said the LORD struck the Ethiopians, *II Chronicles 14:12;* very similar to how he struck Israel before the city.

It also tells us King Baasha provoked the LORD to intense anger, and the LORD dealt with him in the same manner as with King Jeroboam. *I Kings 16:2-5;* King Ahab also provoked the LORD to anger, and was dealt with in the same manner as King Jeroboam. *I Kings 16:33;*

Famines in the Days of Elijah and Elisha

It is time to explore the amazing events of the days of the prophets Elijah and Elisha. There were some truly amazing miracles performed in Israel during the tenure of these two great prophets. Many of these remarkable miracles were performed with attendant celestial wonders and it is necessary to compare these with the effects of the remnant.

It will start with a discussion of the abundant famines in the days of these two potent prophets. Up to now, in the study of the famines brought about by this remnant, there were four major famines. One was in the days of Abraham, one in the days of Isaac, One when the Israelites went to Egypt under Joseph, and one in the days of King David. It is clear each of these previous famines could be a result of the remnant interacting with the Earth.

In the days of Elijah and Elisha, there were four famines of similar magnitude and from similar cause. They are described in scripture as follows:

I Kings 18:2; where the famine was in the days of Elijah and covered all of Samaria,

II Kings 4:38; where a famine occurred in the days of Elisha, and covered the land of Israel,

II Kings 6:24; where a famine broke out in the days of Elisha, and took place in Samaria, and

II Kings 8:1; where a famine erupted in the days of Elisha, in the land of Israel.

As described in the Bible concerning the ministries of these two prophets, these famines were accompanied by many fierce celestial events.

Elijah and the Prophets of Baal

One of the first events, the most major event in the days of Elijah, was the one mentioned above accompanied by a crushing plague. The Bible declares there had been no rainfall for three years, *I Kings 17:1;* and Elijah was diligently looking for the rain, alerting his servants to keep watch for any clouds. During this time, Elijah performed the astounding miracle when he raised the widow's son from the dead. *I Kings 17:17-24*;

Then it mentions the great famine in Samaria, *I Kings 18:2;* followed by the test of gods on the mountaintop with the four hundred and fifty prophets of Baal and the four hundred prophets of Asherah.

Elijah, to prove who was God in Israel, gathered four hundred and fifty of the prophets of Baal, and four hundred of the prophets of Asherah, and set up a sacrifice upon an altar and bade them to call to their gods to come down and consume the sacrifice. *I Kings 18:19-40;* They called all day with loud and passionate lament, even to the point of cutting themselves to stir their gods. Nothing happened.

Then Elijah, to really show the power of the true God, ordered the same setup as the other animal sacrifice, only to his he added the sacrifice be drenched with copious amounts of water. After they finished drowning the sacrifice with water, the ditches around the altar overflowed the area. Then Elijah called unto the LORD to consume his sacrifice and immediately the fire of the LORD fell and consumed the sacrifice, the wood, the stones, the dust of the Earth and even licked up all the water in the trench. *I Kings 18:38*;

Immediately afterward, the enduring drought ended with a robust deluge *I Kings 18:41-46;* and the formidable famine was over. It says the sky was black with obscuring clouds from this commanding event. *I Kings 18:45*;

This was possibly a manifestation of the remnant, raining down a tremendously hot part of itself to the Earth upon the sacrifice. This chunk of rock landed on the sacrifice, at the direction of the LORD and consumed up the whole sacrifice.

The Still Small Voice

Right after the miracle on the mount and the slaughtering of the prophets of Baal and Asherah, Queen Jezebel swore to get revenge upon Elijah and have him killed. At this report, Elijah became fearfully afraid and escaped deep into the wilderness and hid in a protective cave. While Elijah was in the cave, the remnant continued to rain down its injurious effects upon this area of the world.

First, there was a great and strong wind, *I Kings 19:11;* which tore into the mountains, *I Kings 19:11;* and broke the rocks in pieces. *I Kings 19:11;* Second, there was an earthquake, *I Kings 19:11;* followed by, third, a great flaming fire *I Kings 19:12*; and finally, fourth, Elijah found the LORD in the still, small voice and he exited the cave. *I Kings 19:12*; This remnant was buffeting the area of the cave where Elijah had taken refuge, and when it ended, Elijah emerged from the cave, reinvigorated, strengthened and emboldened by his communion with the LORD in the still small voice.

The Fire on the Captains and their Fifties

When Elijah was ready to depart the Earth and be taken up to God in the chariot of Israel, which could have been another representation of the Agent of Mass

Destruction, there were many events occurring explainable by the remnant. Just before he was ready to leave the Earth.

The king sent a captain with his fifty fighting men out to get Elijah and bring him in before the king. Elijah called down fire from Heaven to consume them all. *II Kings 1:10;* When the king heard this, he sent another captain and fifty, and again Elijah called fire down and consumed them. *II Kings 1:11;*

The Parting of the Jordan River for Elijah and Elisha

When the time came for Elijah to depart, many things of a celestial magnitude happened.

First, was the parting of the Jordan River for Elijah and Elisha when Elijah struck the water with his mantle. *II Kings 2:8;*

Secon,d when the river parted for Elisha when he returned from across the river shortly thereafter.

As previously shown, many times before, parting of the waters, rivers or seas, is a manifestation of a very close encounter with this remnant. To further exemplify this remnant was in the vicinity, read of when Elijah was eventually taken up. Elisha left us a detailed description of these notable events.

Third, it tells us, Elijah was taken up to Heaven in a whirlwind, and the Chariot of Israel appeared and took him up. *II Kings 2:1-11;* A cyclonic whirlwind and a flaming horse drawn chariot are obvious manifestations of the remnant of the Agent of Mass Destruction.

The Parting of the Jordan River for Elisha

Then when all those events transpired, Elisha took up the mantle of Elijah, and went back to the Promised

Land, again striking the Jordan River with the mantle, and the Jordan River parted again. ***II Kings 2:14;*** The remnant was still above this area, working its gravity defying ways.

And fourth, shortly thereafter, Elisha made the waters of Jericho good again by sprinkling them with salt. ***II Kings 2:21;*** Previously, something had turned these waters bitter, something affected the drinking water.

The destruction of Sodom and Gomorrah was accompanied by the eruption of much salt in the area, making the resultant Dead Sea the most desolate place on the Earth. Here, Elisha used salt to restore the waters.

From Elisha to Jesus

The Battle of King Ben-Hadad and King Ahab

In the days of King Ahab, there was a battle between the armies of King Ahab and those of King Ben-Hadad. As this battle ensued, there was a great and magnificent manifestation of the remnant. It tells us God would deliver King Ben-Hadad to King Ahab. He did this on the surrounding hills, ***I Kings 20:19-20;*** and in the valleys, ***I Kings 20:28;*** and one hundred thousand men died in one single day. ***I Kings 20:29;*** This description reminds me of a later event when the LORD destroys the host of the Assyrian army of King Sennacherib, discussed later.

Then it states there was an earthquake in the city, where twenty seven thousand died when a wall collapsed upon them. ***I Kings 20:30;***

The Death of King Ahab

The death of King Ahab was predicted beforehand by Micaiah, the prophet of the LORD. ***I Kings 22:17;*** He said he saw Israel scattered on the hillsides. *I Kings 22:17; II Chronicles 18:16;* He also saw the LORD on his throne, *I Kings 22:19; II Chronicles18:18;* and the Host of Heaven (which was previously established to be the planets and moons, including the remnant) on the LORD's right side and left side. *I Kings 22:19; II Chronicles 18:18;* Then he saw the death of King Ahab, even though King Ahab secretly hid himself among his troops. *I Kings 22:34-40;*

Then it says the LORD turned away the Syrians from fighting with King Jehoshaphat, *II Chronicles18:31;* as the Sun was going down, *I Kings 22:36; II Chronicles 18:34;* and the Israelites were thoroughly scattered. *I Kings 22:36;*

These are convincing references to the remnant, as previously shown.

The Battle of Moab and Ammon against King Jehoshaphat

Then the children of Moab and Ammon came against King Jehoshaphat. *II Chronicles 20:1-26;* It informs us the LORD caused a confusion in the enemy ranks, *II Chronicles 20:23;* and every man's sword was against his neighbor, *II Chronicles 20:23;* as in the days when the Philistines tried to return the Ark of the Covenant to Israel.

Again another apparent reference to the remnant.

The Battle of Moab against the Three Kings

During this battle of Moab against the three kings, *II Kings 3:20-26;* there is a graphic description of the remnant wreaking havoc on the Earth.

First, the valley was filled with overflowing water. ***II Kings 3:17-20;***

Second, the undrinkable water appeared as sanguine blood. ***II Kings 3:22-23;***

Third, all the springs of the water were stopped up and no longer flowed. ***II Kings 3:19; 3:25;***

Fourth, all the trees of the land were laid waste. ***II Kings 3:19; 3:25;*** Thus reminds one of the area in Siberia (Tunguska) where a meteor hit devastating thousands of trees in the area, burning them bare and laying them flat in a radiating pattern from the source, completely laying them waste.

Fifth, the good lands were ruined and filled with stones. ***II Kings 3:19; 3:25;*** After the trees were laid waste, stones rained down and filled the whole valley, there were many and they were dispersed over the whole area. In Tunguska, scientists also found numerous stones in the area of the explosion embedded in the surrounding swamps.

And sixth, shortly thereafter, the city of Jerusalem was hit with another epidemic of plague. ***II Chronicles 21:14-15; 21:18-19;***

This description of events are good references to the inauspicious influences of the remnant of the Agent of Mass Destruction.

Elisha and the Syrian Host

In the days of Elisha, there was an incident bearing comparison with the remnant. The Syrian host surrounded Elisha and his servant. ***II Kings 6:13-23;*** The servant was afraid and Elisha prayed for God to open his eyes, and he saw the mountain full of the horses and chariots of fire. ***II Kings 6:17;*** Then the Syrians were all blinded, ***II Kings 6:18;*** as if by a flash of a nuclear warhead. Then was the great famine in Samaria, the first famine in the days of Elisha described above. *II Kings 6:25;*

The Syrian Host were besieging the city of Jerusalem, but the LORD fought the battle, He caused the Syrians to hear the sound of many onrushing chariots, the sound of many galloping horses and the clamoring sound of a great approaching army, *II Kings 7:6;* and they hurriedly fled their camp at twilight. *II Kings 7:7;* The next morning, lepers went out of the city to find provisions and discovered the whole camp of the enemy host abandoned and empty. *II Kings 7:5;*

The great and pervasive famine in the land of Syria lasted seven years, the same length of the famine in the days of Joseph. *II Kings 8:1;* The sounds they heard were being caused by another close encounter with the remnant, which resulted in this sore famine of such an encompassing severity.

The Sundial of King Ahaz Returning Ten Degrees (10°)

Another event which occurred was the return by ten degrees (10°) of the sundial of King Ahaz in the hazardous days of King Hezekiah and the prophet Isaiah. This miracle could most probably have been caused by the remnant of the Agent of Mass Destruction.

It starts when the LORD struck down King Uzziah (Azariah) with leprosy for the remainder of his life, *II Kings 15:5; II Chronicles 26:19;* a plague associated with the remnant.

It says King Ahaz worshipped the Host of Heaven, *II Kings 21:3;* and made his children pass through the fire, *II Kings 16:3; II Chronicles 28:3;* which means he offered them as a human sacrifice, which the LORD commanded them not to do. In fact, it avers all Israel worshipped the

Host of Heaven, ***II Kings 17:16;*** and caused their children to pass through the fire. ***II Kings 17:17;***

There was, likewise, some disordering of the length of the day in the long and tumultuous reign of King Ahaz, and he commissioned a sundial to be built to correctly realign the time of day in accordance with the changes in the orbit of the Earth and its rotation and revolution. This sundial was a result of the disturbances in the celestial sphere, many of which were enumerated here.

When Pope Gregory XIII *CE 1502-1585* revised the calendar in October of *CE 1582*, he was merely revising days from the long overdue revision necessitated by the omission of the correction for its accuracy against the tropical year. The differences between the two calendars was around two one thousandths of one percent (0.002%) and required the addition of 13 days at the outset of the Gregorian calendar.

King Ahaz's sundial was intended to revise the length of the day and the recording of the years after a major disruption of the Earth's rotation and revolution in his days. This was no mere intercalary correction.

One day, Isaiah, a prophet of Israel came to King Hezekiah as the LORD commanded him, and told King Hezekiah to get his house in order, because he was going to die. ***II Kings 20:1;*** King Hezekiah ruled as a righteous king and appealed to the LORD, on those grounds, for additional years of life. ***II Kings 20:2-3;*** God answered his prayer and sent Isaiah back to him to tell him he was granted more time. Isaiah returned to him and told him he would be completely healed and have an additional fifteen years of life. ***II Kings 20:4-7;*** However, King Hezekiah requested a sign. ***II Kings 20:8;*** The sign agreed upon was for the sundial of King Ahaz to return back in its course by ten degrees (10°) and the LORD accomplished this. ***II Kings 20:9-11;***

For this to have happened, the Earth would need to go backward on its own axis by a total of forty minutes. This was probably another encounter with the Agent of Mass Destruction, which was lingering in Earth's orbit for most of the history of Israel in the land, coming and going.

This forty minutes also changed the way men kept time and probably further confused the calendar in the days of King Hezekiah, requiring Emperor Julius Caesar *100-44 BCE* to revise it once again in *46 BCE.*

Shortly after this, a major event occurred on Earth which may also be evidence of the irreversible Agent of Mass Destruction.

The Destruction of King Sennacherib's Army

In the days of King Hezekiah, a truly amazing event occurred associated with the remnant. In the days of King Hezekiah, the ten northern tribes had already been carried away into captivity by the Assyrians. The land of Judah, to the south was under attack by this Assyrian Host, and they were besieging the city of Jerusalem, but the city was holding out against the siege. King Hezekiah prayed to the LORD for swift deliverance and the LORD delivered him and Judah in a mighty way.

The prophet Isaiah told King Hezekiah the deliverance would come and King Hezekiah wanted a sign. God gave King Hezekiah a choice, the sundial of King Ahaz could go forward or return ten degrees (10°). ***II Kings 20:9;*** King Hezekiah said, make it go backward and it was so. ***II Kings 20:11; Isaiah 38:8;***

For the sundial of King Ahaz to return ten degrees (10°) on the dial, the Earth's axial rotation had to be slowed, stopped and sent backward forty minutes. It reflects the power of the remnant to affect the Earth's orbit and rotation. This reminds us of the time when Joshua told the Sun and Moon to stand still during the slaughter of the Amorites in the valley of Gibeon.

Shortly thereafter, the blast from the LORD destroyed one hundred eighty five thousand troops of the Assyrian army and delivered Jerusalem from the siege. *II Kings 19:35; II Chronicles 32:21; Isaiah 37:36;* Then another plague hit the city, and it states King Hezekiah almost died from a bothersome boil, *II Kings 20:1; 20:7; II Chronicles 32:24;* another recurrent manifestation of major encounters with the remnant.

The Worship of the Host of Heaven

A number of Israel and Judah's kings are said to have worshipped the Host of Heaven. Along with those already mentioned above, there was King Manasseh, who provoked the LORD to anger, *II Kings 21:6; II Chronicles 33:6;* worshipped the Host of Heaven *II Kings 21:3-5; II Chronicles 33:3-5;* and made his children to pass through the fire. *II Kings 21:6; II Chronicles 33:6;* King Josiah was said to have ritually cleansed the temple and all of the implements used in the worship of the Host of Heaven, *II Kings 23:4;* and King Josiah removed the priests and officials who participated in these abhorrent and prohibited activities and ceremonies, *II Kings 23:5;* and here is where the Host of Heaven are identified as the Sun, the Moon and the stars. *II Kings 23:5;* He also burned the stultified and tainted altar where the people made their children pass through the fire. *II Kings 23:10;* He removed the sculpted and carved idolatrous horses dedicated to the Sun, *II Kings*

23:1; and he burned the iconic Chariots of the Sun. ***II Kings 23:11;***

The Siege of Jerusalem by the Babylonians

During the siege of Jerusalem by the Babylonians, there were again a number of events relating to the remnant. There was a famine in Jerusalem in the days of King Zedekiah. ***II Kings 25:3;*** It explains the wall was broken through, ***II Kings 25:4; Nehemiah 1:3; 2:3; 2:13; 2:17;*** and by implication it was done, not by the army, but by another event. The Chaldeans were still encamped around Jerusalem. ***II Kings 25:4;*** It further declares a fierce fiery fire burned all the gates of Jerusalem, ***Nehemiah 1:3; 2:3; 2:13; 2:17;*** it even burned the stones of the wall. ***Nehemiah 4:2;*** This was a tremendously destructive fire.

These were obvious references to an event of celestial proportions, as were many similar events in history as previously explained.

The Days of Queen Esther

In the days of Queen Esther, the Bible tells us, the city of Shushan was in a state of utter confusion. ***Esther 3:15;*** This confusion was being caused by an outside event, an event of celestial proportions. It was the effects of the remnant.

Famine in the Days of the Building of the Second Temple

The final famine of great proportion spoken of in the scriptures was the one in the days of Nehemiah, when he returned to the Promised Land from the Babylonian captivity and by the edict of King Cyrus, the Persian, an edict allowing them to rebuild the broken down temple. ***Nehemiah 5:3;*** Like all other famines described in scripture and history, this one was probably a manifestation of the remnant.

From Jesus to Today

The Chaldeans

One major event with far reaching implications for the people of the Middle East, was the destruction of the army of King Sennacherib besieging Jerusalem at the height of the Assyrian power. ***II Kings 19:35; II Chronicles 32:21; Isaiah 37:36;*** It spelled the ultimate downfall of the Assyrian kingdom, because the Assyrians severely diminished as a power afterward and the Babylonians rose thereafter and flourished.

The Babylonians, and the Assyrians, knew the cause of the destruction of King Sennacherib's army. They knew it was as a result of the remnant. They made a science of tracking the celestial bodies, especially the remnant. The Chaldeans became this class of magicians who understood the movement of the Host of Heaven, and they studied it carefully. They did not want to be caught unawares when it again came into Earth's vicinity to trouble mankind.

Throughout their illustrious history, they suffered many instances when they were called upon to give answers none of the other officials of the kingdom could offer. When King Nebuchadnezzar dreamed his famous statue dream, ***Daniel Two;*** he called on the Chaldeans to give the proper interpretation, which they were unable to perform. It fell to a humble prophet of Yahweh, Daniel, to give an accurate interpretation.

To understand them as a special class and not just your run of the mill magicians in Babylon, the Bible informs us they were called by King Nebuchadnezzar, as well as the magicians, the astrologers, and the sorcerers. ***Daniel 2:2;*** It would not have identified them as a different group from the magicians, if they indeed were just simple ordinary court magicians. To show their primary importance as acute diviners, they responded first of all the classes summoned. ***Daniel 2:4;*** It was their understanding of the Host of Heaven making them rightly suited to be the ones identifying the star in the East heralding the coming of the King of the Jews, the Star of Bethlehem.

The Star of Bethlehem

In order to see the true extraterrestrial nature of the Star of Bethlehem, it is necessary to understand how the gifted knowledgeable wise men, the magi, (Chaldeans) from the east, came to Jerusalem to find the heralded King of the Jews. They saw the star in the east first, ***Matthew 2:2;*** and they then travelled west to Jerusalem. ***Matthew 2:1;*** They correctly recognized the star as the portent of the birth of the King of the Jews. ***Matthew 2:2;*** They were in Babylon, and they looked east, toward China, and saw the bright and unusual star, and knew immediately, through their years of studying the Host of Heaven, it indicated the arrival of the King of the Jews, to the west of them. The impression is they followed the star to Jerusalem and thereby knew it must have heralded the birth of the King of the Jews, when they arrived at Jerusalem. This is not what it says. It says they saw the star in the east, and went west.

The reason they knew it as the star heralding the birth of the King of the Jews, was because, since the destruction of the army of Sennacherib, and perhaps for this reason, they were studying the remnant. And now it reappeared, and they immediately knew the timely auspicious reappearance of the remnant portended something of great significance for the Jews, deliverance, as were all the other manifestations of the remnant in the past. They associated the remnant with the Jews, with good things for the Jews.

When they came to Jerusalem, the star was not visible to them, further proof the star was not guiding them. They went in to ask King Herod where the child was supposed to be born. *Matthew 2:2;* When they received their answer, they departed King Herod for Bethlehem. *Matthew 2:6;* The momentous star again appeared to them and they rejoiced again to see the star. *Matthew 2:9-10;* The star came to where the child was, and they were able to find the child. *Matthew 2:10;* The star which was in the east when they first observed it from their base in Babylon, now appeared in the north, north of Jerusalem, Bethlehem is south of Jerusalem, but Nazareth is north of Jerusalem. Jesus was already back in Nazareth. *Luke 2:39;*

This star was moving across the sky.

First it was in the east, and the reading of the sign caused the Chaldeans (wise men) to go west to seek him who was to be born, King of the Jews. *Matthew 2:2;* The star was not in the sky when they were talking to King Herod, because he told them to go and search for the child in Bethlehem, south of Jerusalem. *Matthew 2:8;* If the star was visible in the sky when they arrived in the Holy Land, they would not have needed to ask King Herod where the child was to be born, they could have just followed the star.

When it reappeared north of Jerusalem, they rejoiced exceedingly, and this indicates it was not visible then, but reappeared and led the wise men to Nazareth. They were headed south from Jerusalem to Bethlehem, where the child had been born in a manger in a barn or a cave. They were headed in the wrong direction, because the child was north of Jerusalem in the house in Nazareth. **Matthew 2:11;**

When the star reappeared, the wise men rejoiced exceedingly because it put them back on track, heading north to where the young child was, **Matthew 2:11;** not south where the child was prophesied to be born, possibly as much as two years earlier, based on the timing of their sighting of the star as told to King Herod, **Matthew 2:7;** and King Herod's later command to slay all the children in Bethlehem, two years old or younger. **Matthew 2:16;** They told King Herod they had seen his star in the east two years prior and this prompted the paranoid King Herod to order the Massacre of the Innocents of two years or younger. It further says they came to the house where the young child was, **Matthew 2:11;** not in a barn, a stall, a cave or a manger.

They were tracking a piece of the shattered debris from the Agent of Mass Destruction. They recognized it as the star heralding the birth of the King of the Jews, because they associated the Agent of Mass Destruction with events of momentous concern for the Jews, being the Agent of the Destruction of the Assyrian army when it assailed and threatened the Jewish capital.

They, through their studies, were aware the fate of the Jewish nation, and almost all of its momentous events, were aided by the Agent of Mass Destruction.

For example, just a few of the many encounters with the remnant are the flood in the days of Noah, the destruction of Sodom and Gomorrah, the Exodus, the conquest of Canaan, the destruction of the prophets of Baal by Elijah, and the destruction of the Assyrian army outside Jerusalem.

Then the three wise men departed Judea another way to their own country. **Matthew 2:12;**

Then King Herod sent his troopers to kill all children two years old and younger in Bethlehem, **Matthew 2:16;** because he thought the child would be in Bethlehem, where he sent the wise men. According to the time he diligently inquired of the wise men, he suffered all the male children two years and under to be slaughtered. **Matthew 2:16;**

Jesus was probably as old as two years when visited by the wise men. And the star could have appeared two years before in the east.

The Crucifixion and Resurrection of Jesus

At the crucifixion of Jesus, there was a major earthquake in Jerusalem. It was severe enough to rend the veil of the Holy of Holies into two pieces, from top to bottom. The Earth quaked and the rocks were rigorously rent.

It also says during the earthquake accompanying the crucifixion, the graves were opened. The inhabitants of those graves did not arise and go into the city until three days later, when Jesus rose from the dead. In the city they were seen by many. **Matthew 27:51-53; "And, behold, the veil of the temple was rent in twain from the top to the bottom; and the Earth did quake, and the rocks rent; And the graves were opened; and many bodies of the saints which slept arose, And came out of the graves after his** *resurrection, and went into the holy city, and appeared unto many."*

This was a result of an encounter with the remnant. The remnant stayed in the vicinity of the orbit above Jerusalem for some days, causing many unusual events.

The Trees of Ireland *CE 364*

In Ireland, there is an isolated island constructed of trees, soil and peat in the midst of a lake by the early Celts in the year *CE 364*. This island was used as a defensive position against the invading Roman legions.

This was a common practical practice among the various Celtic tribes of the British Isles as an adequate protection against hostile invaders. They would build islands in the middle of lakes, and inhabit these natural and formidable fortifications for the safety of the tribe and for advantageous strategic strongholds against the enemy.

On one of these surviving man made islands, the water-logged timbers used as the foundation are still relatively intact having been preserved by the peat. Scientists have analyzed these timbers using the science of dendrochronology. The tree rings of these analyzed timbers reveal there was a major celestial encounter with the remnant at this approximate time. Scientists call it an encounter with an asteroid. All encounters with asteroids are merely encounters with remnants of the Agent of Mass Destruction.

The Justinian Plague *CE 567*

Between *CE 540-596*, a plague of enormous proportions swept the Eastern Roman Empire, especially virulent at Constantinople. This was during the reign of

Emperor Justinian *CE ca 482-565* and is therefore referred to as Justinian's Plague.

This agonizing plague produced a devastating effect on the Eastern Roman Empire, killing vast numbers in the population over a wide area now comprising much of Europe, European Russia, Asia Minor and the Fertile Crescent.

At the same time this sweeping plague was ravaging the Eastern Roman Empire, there were many reports of disturbances in the Heavens. The Gallo-Roman historian, Gregory of Tours *CE 538–594* reports in *CE 567*, over Auvergne France, the sky above the city was on fire, and he also reported in Rome a great fiery dragon was seen over Rome, an apparition seen burning across the length of the sky and ending up plunging into the Mediterranean Sea off the coast.

In *CE 590*, around Avignon France, it was noted the midnight was as bright as the day.

It is obvious the remnant was still hovering around and wreaking havoc on man.

The Battle of Hastings *CE 1066*

In *CE 1066*, the Norman Duke William I (the Conqueror) *CE 1058-1087* crossed the English Channel and invaded the realm of Harold II Godwinson, *CE 1022-1066* the King of the Britons.

The Bayeux Tapestry, *ca. CE 1470s* hanging in the Bateaux Cathedral in the town of Bayeux, France was made to commemorate the event. The tapestry, in one of its many panels, shows what appears to be a streaking comet in the skies above the battle and the invading Duke William mounted on his charging stallion waving his helmet in the air to allow his troops to see his face and know he was yet surviving.

Scientists have surmised this comet may have been an early representation of what would later become Halley's Comet. It could easily have been another piece of the remnant entering the upper atmosphere of the Earth above Britain.

The Plague of *CE 1118*

In *CE 1117* a large and incredible comet was seen passing north over Europe toward Asia. The light of the Moon appeared blood red after this comet passed, which would indicate there was a lot of iron oxide dust expelled into the atmosphere of the Earth by this passing comet. It reminds one of the object invading Egypt during the days of the Exodus.

Then in *CE 1118*, a light appeared over Europe more brilliant than the Sun, which was followed by extreme cold, a great famine and prolific plague.

These are classical descriptions of encounters with the remnant.

The Little Ice Age *CE 1350-1850*

From *CE 1350-1850*, an epochal event occurred worldwide which is now commonly referred to as the Little Ice Age. The mean temperatures dropped significantly all over the world during these five hundred years. Scientists are still at a loss to explain this freezing phenomena.

This thesis proposes it was a result of another encounter with the remnant.

In the event which manifested the plague of *CE 1118* there was a period of extreme cold immediately following for a short period which seems to be another by-product of the Agent of Mass Destruction and this Little Ice Age was one of extreme duration.

This Little Ice Age lasted for about five hundred years and then the daily temperatures again began to normalize. The fact this remnant could affect temperatures is another aspect of the celestial marauder.

The Bubonic Plague of the Fourteenth Century *CE 1348*

The Bubonic Plague in Europe during most of the Fourteenth Century, *CE 1333–1400*, also called the Great Plague and the Black Death was the worst and most devastating plague to ever strike mankind. It started in Asia, somewhere in China and depopulated much of China before striking Europe, however, its effects in Europe were particularly noteworthy. In Europe it was said to have killed up to half of the continent's population. It was a ravaging, appalling and very contagious plague, which was spread from person to person and was airborne. It caused painful swelling of the lymph nodes leaving black oozing sores (buboes) and liberating death usually occurred within three to seven days from onset. There was little an unenlightened and unprepared medical establishment could do to retard and halt its deadly and insidious spread.

Accompanying the plague were also many descriptions of celestial events.

In Asia, it is reliably reported the plague, which began there was preceded by fierce storms, earthquakes, numerous falling meteors, all combining to level trees and destroy the fertile places. There were also many comets seen in the Heavens.

This description resembles the calamity which struck Heshbon, in the days of the Conquest of Canaan by the children of Israel. This could definitely be the result of the remnant.

Johannes Nohl reports between *CE 1298-1314*, seven large comets were seen over much of Europe, one was unusually black (an asteroid, a part of the remnant). Shortly thereafter, the Black Plague started in Europe in earnest.

This is a reference to the remnant visiting death upon Europe and Asia during these bleak years.

In *CE 1347*, a "column of fire" was observed over the anti-pope's palace at Avignon, France. This reminds one of the column of fire and column of cloud accompanying the Israelites in the wilderness.

Also, in *CE 1347,* a ball of fire was observed for an extended period of time over the city of Paris. Shortly thereafter, the plague broke out in earnest in Paris.

The Comet over Arabia *CE 1479*

In *CE 1479*, over the Arabian Peninsula, a comet was seen streaking across the sky. This was a meteorite, since comets are not usually described as streaking across the sky and are rather described as being seen in the Heavens. This was another piece of the remnant.

Numerous Comets Seen Between *CE 1500-1650*

The major part of the remnant broke up into numerous smaller chunks during this time because the reports of comets and meteors and stones falling from the sky were many more numerous during this auspicious century and a half.

Between the years *CE 1500-1550* there were at least twenty six 'comets' seen over Europe. Between *CE 1550-*

1600 sixteen more 'comets' were seen. These included one in *CE 1568* which caused a plague in Austria and one in *CE 1582* accompanied by a malicious plague in the Lowlands. In between the years *CE 1600-1650* there were nine more 'comets' observed, including the big comet of *CE 1606* which was followed by a great and harmful plague.

These were all a result of the breakup of this remnant, and the resultant rain of rocks from the remnant to the Earth.

The Maunder Minimum *CE 1611*

In *CE 1904* a man and wife team, British astronomers, Annie Russell Maunder *CE 1868–1947* and her astronomer husband Edward Walter Maunder *CE 1861-1928* discovered there were no sunspots on the Sun for most of the *CE seventeenth century*. This lasted for a duration of seventy years from *CE 1645–1715*, which is seven cycles of sunspots on the Sun. During this time, there were no sunspots and scientists do not understand a reason for this deficiency. After these seven missing cycles, sunspots returned to the surface of the Sun and have been regular ever since.

Perhaps it was the Agent of Mass Destruction.

The Great Plague of London *CE 1664*

Another plague visited London in the year *CE 1664*.

This expansive thesis has already spent considerable time identifying the remnant as the cause of these various and pestilential plagues and when the plague occurs anywhere, it could possibly be pointing to an encounter with the remnant.

The Great Earthquake of Lisbon *CE 1755*

On the morning of *1 November, CE 1755*, a great and damaging earthquake hit the city of Lisbon, Portugal laying the city and surrounding area to waste. It was estimated to be between 8.5 and 9.0 on the Moment Magnitude Scale (MMS) which replaced the Richter Scale in the *CE 1970s*.

This earthquake was massive, it affected most of the Iberian Peninsula and into Morocco, across the Straits of Gibraltar. This is a truly vast extent for an earthquake to affect. The epicenter was at Lisbon in Portugal. Approximately eighty five percent (85%) of all the buildings in Lisbon were destroyed by either the earthquake, the ensuing fire-storm or the enormous tsunami following.

The death toll lies somewhere between forty and fifty thousand by all educated estimates, since records are rare and those which do exist are vague and jumbled, mixing up the *1 November* earthquake with other earthquakes and aftershocks in the area around *18-19 November* of the same year.

Earthquakes have been shown previously to be a possible result of the remnant of the Agent of Mass Destruction, and this earthquake was mass destruction on stark display.

The Rain of Stones over France *CE 1860*

In *CE 1860*, there was a tremendous fall of stones from the sky over Bayonne, France. This was when science came to realize meteors were falling from space onto the Earth. This fall of rocks was one in many falls from the breakup of the remnant.

The Tunguska Explosion *CE 1908*

Scientists now are fairly certain the explosion, which occurred over the Tunguska area of Siberia in *June CE 1908*, was the result of an encounter with an asteroid or comet. It was another encounter with some of the last remaining pieces of the remnant.

Today scientists are trying to track all the asteroids in Earth's vicinity. These Earth crossing asteroids are remnants of the fateful collision which first happened perhaps millions of years ago. The remnant, over the millennia, through numerous encounters with the Earth and the Moon, was reduced to rubble and the Earth crossing asteroids are the remains. It has been a bane to mankind throughout history.

Conclusion

This brings a close to this thesis.

There are three sections to this thesis.

The first attempting to reconcile the Holy Bible to scientific findings and showing there is no dichotomy between the two.

The second part lays down an extensive scientific treatise showing the role of catastrophism in the pre-historical records of the Earth.

The third part explains how man's historical record, especially the Holy Bible, describes this catastrophism and man's attempts to describe in his limited way what he was seeing and experiencing.

This thesis conjectures the Agent of Mass Destruction which made a journey through the Solar System in the dateless past, leaving devastation in its wake and which scientists are currently trying to locate beyond the orbit of Pluto.

If scientists would focus their search closer to home, they find the last major piece of this marauder in the inner Solar System, in the next closest planet in toward the Sun, the planet Venus, the only planet rotating retrograde.

That is a topic for another work.

THE END

About the Author

Reggie Frasier is an Award willing author with numerous titles making up his portfolio.

He was educated in Miami, Florida in the late 1960's and graduated in 1970 from Miami Killian Senior High School in that city. After high school he enlisted in the United States Air Force and served with distinction until the end of the Vietnam War, when he was honorably discharged and moved to Atlanta, Georgia where he attended college at Georgia State University in 1978. After he finished his curriculum at Georgia State University in 1982 he worked a variety of careers until he settled into his career with the United States government at the Department of the Treasury. He retired from this career in 2008 and began "working seriously" on his writings which he had half-heartedly been pursuing since his college days.

He was ordained as a minister of the gospel in 2004 and has spent most of his adult life studying God's Word and many of his books concern these studies.

He met and married his wife in 1975 before he started college and she was instrumental in encouraging him in that pursuit. They were blessed with a son in 1993. He lives with his wife and son just outside Atlanta, Georgia.

He is somewhat of a medical miracle because of the number of times he has faced life threatening situations. At ten years old, he contacted Scarlet Fever which was extremely rare and which he survived, When he was thirty six years old, he suffered a heart attack (an MI) which he survived. At age sixty one he was diagnosed with colon cancer which he survived and today is cancer free. All these "miracles" influenced and contribute to his writing.

www.ingramcontent.com/pod-product-compliance
Lightning Source LLC
Chambersburg PA
CBHW070026210526
45170CB00012B/52